D0119653

50 ideas

you really need to know

earth

Martin Redfern

Quercus

Contents

Introduction

We inhabit a wonderful planet. We are fortunate if we can take the time to marvel at its beauty, gaze in awe at its majesty and be thankful for the gifts it brings. But for most of our busy lives, we scurry about on the surface and forget two important dimensions: depth and time. In this book, I hope to remind us of those forgotten dimensions.

Consider for a moment what lies beneath your feet – not just the familiar earth and rock of the surface layers, but deep down. As close to you now as the distance many of us commute every day lies a place no one has visited and conditions of temperature and pressure we can scarcely imagine. Travel less than the distance of a transatlantic flight and you would find yourself in an incandescent world of molten metal. The Earth is not just sitting there like a block of concrete waiting for us to walk all over it. It is a living, dynamic planet. Solid rocks are on the move as continents drift, volcanoes erupt and the vast, deep mantle slowly churns. Neither are the rocks below the surface immune from the processes going on above them. Water, air and life itself are in constant dynamic interaction with the geology. Without oceans, we would not have continents. Without life, we would not have our atmosphere or a climate in which we could live. The natural cycles of our planet have supported life for billions of years. We interfere with them at our peril.

The other dimension opened up through understanding the processes at work in our planet is the dimension of time. Not just time as in lunchtime or even lifetime, but deep time. It takes a radical change in the way we think in order to get our minds around time when it is measured in tens and hundreds of millions of years, but that is what we must do if we are to understand our home. Once we have made that change, we start to realize that everyday processes extended over deep time can build and destroy mountain ranges, open up oceans and split continents apart. Deep time can create new species or drive them to become extinct. Our human existence scarcely registers as one tick on the clock face of deep time and yet we have already changed the planet beyond recognition. Perhaps if we come to understand it better, we will treat our world more kindly.

01 Birth of Earth

We are all made of stardust. The primordial hydrogen and helium created in the Big Bang 13.7 billion years ago has been cooked in the nuclear furnaces of generations of stars to produce the carbon, oxygen and nitrogen of our bodies; and the silicon, aluminium, magnesium, iron and all the rest of the elements that make up our planet.

Stardust memories Stars shed their outer layers towards the end of their lives. Massive stars can no longer support their own weight and they collapse, triggering a supernova explosion that scatters their ashes in great clouds of dust and molecules. It was out of such a cloud that our solar system was born. Every molecule in your body contains elements that were cooked in stars. Every atom of gold in the ring on your finger was created in a supernova.

The presence of decay products of short-lived radioactive isotopes in ancient meteorites suggests that these elements had their origins in a nearby supernova explosion not long before the solar system formed. Indeed, it may have been such an explosion that triggered the initial collapse of the solar nebula.

Accretion As the gas and dust were drawn towards the centre where the Sun would eventually form, angular momentum in the gently rotating nebula would have flattened the material out into a disc. For a long time that was just theory, but now powerful telescopes can see it happening in other stellar nurseries. For example, the star Beta Pictoris has a clearly

timeline

4.6 Ga	4.567 Ga	4.54 Ga	4.527 Ga
Possible supernova explosion; solar nebula begins to contract	Age of chondrules in meteorites, the first solids in the solar system	Proto-Earth reaches the size where melting begins and the core separates out	Formation of the Moon

catch a falling star

The first solid grains to form in the young solar nebula were chondrules. These are roughly spherical grains of silicate rock, ranging from a fraction of a millimetre to a centimetre in diameter. They appear to have formed as molten droplets when silicate dust was heated to around 1,500 degrees Celsius, presumably close to the new Sun or perhaps by radioactivity. They are found today in about 80 per cent of all the meteorites that land on Earth and can be dated with amazing accuracy. At 4,567 million years old (give or take 0.5 million years), they are the oldest things in the solar system.

visible disc of dust and rocky grains around it, which could be forming into planets right now. The detection of so-called exoplanets around over a thousand other stars suggests that planetary formation frequently accompanies starbirth.

It is generally agreed that the planets in our solar system built up by a process called accretion, with small grains bumping into one another and collecting together. The first part of that process is the hardest to understand, as there would be little gravity to hold the clumps together and collisions would tend to break them up again. It is possible that concentrations of grains may behave like a kinetic liquid, holding together and only occasionally gaining enough energy to 'splash' out of the cluster. If the relative velocities of the grains were slow enough, they would begin to stick together. Once they had reached the size of a few metres in diameter, gravity would take over, drawing more and more material together.

4.42 Ga	4.404 Ga	4.28 Ga	3.85 Ga
Oldest mineral grain from Apollo lunar samples	Oldest mineral grain on Earth. Possible evidence of water	Oldest surviving rock on Earth, possibly from a deep sea vent, from Hudson Bay in Canada	Oldest surviving sediments from Greenland

Stellar alchemy

Stars are nuclear furnaces. Like hydrogen bombs, they convert the most abundant elements in the universe, hydrogen and helium, into heavier elements, in the process releasing the energy that makes stars shine. Ordinary stars produce the elements of life – including carbon, nitrogen, oxygen – and those that make up the bulk of the Earth – such as sodium, potassium, calcium, aluminium and silicon. As a star ages, it sheds these elements off into space. Some stars produce so much carbon that they are surrounded by clouds of soot. The endpoint in that sequence is iron. To make anything heavier requires more energy than it releases. So, when the heart of a massive star has turned to iron, nuclear fusion stops. The star can no longer support its great mass and it collapses, triggering an incredible explosion that blows the star apart and creates the full range of heavy elements right the way down to uranium.

Separation Gravitational energy, the heat of radioactive decay and the energy released by the impacts of collisions would have led to melting, eventually enabling the heaviest elements such as iron and nickel to sink down and form a core in a body that is now roughly spherical and perhaps tens or hundreds of kilometres in diameter. That body would continue to mop up remaining dust and larger fragments to form a smaller number of protoplanets. Collisions between these would be less frequent but more violent.

The wind from the Sun The formation of the Sun probably only took about 10,000 years, by which point enough matter had been squashed together so that it reached the temperatures needed for nuclear fusion to begin and the Sun to start to shine. That resulted in a strong solar wind of particles blowing out through the young solar system. It would have stripped away any early atmosphere of hydrogen and helium from the Earth, leaving the more resistant rocks of the planet. The bulk of the gas collected further out to form the giant gas planets, Jupiter and Saturn. Volatile material such as methane and water condensed even further out, forming the icy bodies of the outer solar system: dwarf planets such as Pluto, ice moons, Kuiper belt objects and comets.

A new planet Our young Earth continued to grow. The interior was probably now mostly molten, with an iron core surrounded by the primitive silicate mantle. Once it had grown to about 40 per cent of its present mass, gravity would have helped it to retain an atmosphere, while a magnetic field generated in the core might have protected it by deflecting solar particles. That first atmosphere probably consisted mostly of nitrogen, carbon dioxide and water vapour.

As we will see over the next few pages, the accretion process continued, culminating in the major impact that formed the Moon. As the young Earth cooled, liquid water could exist on the surface. Some of the water vapour may have been produced by the planet as volcanic gases, but much of it probably fell to Earth in icy comets, along with the rocky material from meteors and asteroids. It is an accretion process that continues in a small way to this day. If you go out on a clear, dark night, you may see shooting stars. These meteors are small grains of solid material burning up in the atmosphere but ultimately finding their way onto our planet. Each is no bigger than a grain of sand or at most a grain of rice, but between them they add up to between 40,000 and 70,000 tonnes every year, continuing the process by which our planet was born.

the condensed idea
Growing planets by accretion

02 Our companion Moon

When our planet was less than 20 million years old, it suffered the most catastrophic event in its existence. Another planet the size of Mars crashed into it at around 30,000 mph! The impact melted the Earth, but it also gave us a companion that has stabilized the seasons and opened the way to life: the Moon.

Over the years, there has been much speculation about the origins of the Moon. Before the theory of continental drift was accepted, some speculated that the Moon had somehow spun off from a bulge where the Pacific Ocean now lies. Others proposed that it had been formed alongside the Earth by a similar process of accretion, or that it was formed elsewhere and captured in passing by the Earth's gravity. But none of these explanations quite fitted with what we know about the Moon's orbit.

A chip off the old block It was only when the Apollo astronauts visited the Moon and brought back rock samples that the truth began to dawn. The Moon rocks had a very similar composition to volcanic basalts and mantle rocks of the Earth. We were made of the same stuff.

Now, with the help of computer simulations, scientists have a pretty good idea what must have happened. Another protoplanet could have formed in a so-called Lagrangian point ahead or behind our planet, so that it was an equal distance from both the Earth and the Sun. If it formed from the same ring of material in the solar nebula, that would explain why it had the

timeline

4.527 Ga	4.42 Ga	4.36 Ga
Probable time of the impact from which the Moon originated	Oldest dated lunar mineral grain	Oldest dated lunar rock sample

same composition as the Earth. As it grew, the orbit became unstable and it ended up on a collision course with Earth. That object has been called Theia, after a Titan of Greek mythology, the mother of Seline, the Moon goddess.

Cosmic crash Travelling at around 10 miles (16 km) per second, Theia would have loomed in the Earth's young sky for several days, getting nearer and nearer. In the end, the impact was all over in a flash. Within seconds, supersonic winds stripped away the Earth's atmosphere. Almost instantly, much of Theia's mantle as well as some of the Earth's was vaporized and flung into space. Most of Theia's dense iron core looped around the Earth and impacted a second time to merge with our own core. The rest swept out into space, dragging incandescent streamers of molten rock behind it. All of that must have happened in about 24 hours. Viewed from a safe distance, it would have been an incredible sight.

Gradually, most of the material fell back to Earth, but enough remained in orbit in an incandescent ring around the Earth's equator. As it cooled, it condensed into particles, which congealed together over the next few decades to form the Moon. Some of the surprises in the composition of Moon rocks returned by the Apollo missions can be explained if those rocks had condensed from silicate vapour in a vacuum.

EUGENE SHOEMAKER 1928–97

Gene Shoemaker (1928–97) was a pioneering lunar geologist. He studied Meteor Crater in Arizona and used it to show that most of the craters on the Moon were caused by impacts, not volcanoes. He hoped to become an astronaut himself, but was disqualified for medical reasons. He nonetheless played an important part in selecting the Apollo landing sites and training the astronauts. Following his death in a car crash, some of his ashes were placed aboard Lunar Prospector and delivered to the Moon in 1999.

4.1–3.9 Ga	3.6 Ga	3.1 Ga
Heavy bombardment created the maria basins	Lunar core freezes. Lunar magnetic field turns off	Last big basalt eruption in the maria basins

Second moon It is possible that not all the ejected material was collected quickly into a single moon. There are suggestions that a second moon, about 1,000 kilometres (621 miles) across, was formed at the same time and continued to orbit the Earth for several million years before eventually merging into our Moon in a relatively gentle impact. If that impact was on what is now the far side of the Moon, that might explain why the crust there is about 50 kilometres (31 miles) thicker than on the near side and why there are differences in composition between the two sides of the Moon.

As the crust of the Moon began to solidify, certain elements would have been left in the molten material sandwiched between the crust and the mantle. These included high quantities of potassium (K), Rare Earth Elements and phosphorus (P), leading this to become known as KREEP-rich magma. Accretion of another small moon onto the far side of our own Moon would have squeezed that molten layer around to the other side, causing the near side of the Moon to be particularly rich in the KREEP elements.

Short days, brilliant nights Theia's glancing blow to the Earth would have made our planet spin faster. Day length following the collision was only about five hours and has been getting steadily longer ever since. The newborn Moon would also have been much closer to the Earth, appearing about 15 times bigger in the sky – a spectacular sight if you had been able to stand on the Earth's glowing, volcanic surface. The Moon's tidal effects would have been far greater than today, though there were no oceans to experience them. But there would have been massive Earth tides in the molten magma beneath the surface, perhaps increasing volcanic activity each time the Moon passed overhead.

> **It suddenly struck me that that tiny pea, pretty and blue, was the Earth. I put up my thumb and shut one eye, and my thumb blotted out the planet Earth. I didn't feel like a giant. I felt very, very small.**
>
> **Neil Armstrong**

Ever since, the Moon has been getting gradually further away as the tides sap its orbital energy.

Prospecting for water

Following the Apollo space programme, there was a long gap in lunar exploration. But more recently, several unmanned craft have returned to the Moon, and one of their priorities is to look for water. Lunar Prospector detected abundant hydrogen around both the lunar poles, leading to suggestions that it could be in the form of water ice in shady craters. In 2009, the US LCROSS probe crashed into a crater near the South Pole, producing a plume of ejecta that, while not as spectacular as anticipated, contained an estimated 155 kg (342 lb) of water ice in fine crystals. The Indian Chandrayaan 1 probe used radar to detect ice beneath the surface near the North Pole. These discoveries could be important, as they could supply rocket fuel for future missions and perhaps water for settlers.

Within a few million years, tidal forces locked the Moon so that one side always points towards the Earth. Laser measurements using reflectors left on the Moon by the Apollo astronauts show that today the Moon is still moving away from us at 3.8 centimetres (1½ in) a year.

Destroyer and protector It is possible that primitive life had already gained a toehold on Earth before the cataclysmic collision. If so, it was completely annihilated and there must have been quite a wait before volcanic eruptions and impacting icy comets replenished the atmosphere and oceans. But it was worth the birth pains and the delay. Without the Moon, not only would we lack the tides, but also the Earth's rotation axis would be unstable – it might have flipped at irregular intervals, perhaps pointing one pole towards the Sun and leaving half the world in darkness. We would also have lost the most beautiful object in the night sky.

the condensed idea
Interplanetary collision

03 Hell on Earth

For the first 700 million years of its life, planet Earth was not a pleasant place to be. This has been called the Hadean eon, named after Hades or hell. It was a time of terrible bombardment by asteroids and constant volcanic eruptions. At times, all or part of the Earth's surface was molten magma; any atmosphere was stripped away and oceans vaporized. And yet it is also the beginning of the world as we know it.

A brief history of the Moon The young solar system was still a dangerous place around 4 billion years ago. As smaller objects merged, impacts became less frequent but more violent. This is the episode known as the late heavy bombardment, which continued until about 3.85 billion years ago. Traces of that bombardment have long since been wiped clean from the face of the Earth, but on the Moon they are still clearly visible.

It was the late heavy bombardment that created the dark patches we see on the face of the Moon today. These are the lunar seas, or maria. No ship has ever sailed them, but they were once liquid – liquid lava. They were caused by huge eruptions of basalt magma into the vast basins created by the bombardment. They offered a relatively flat surface on which the first Apollo landers could touch down. The samples they returned were ancient by Earth's standards. Even the youngest Moon rocks dated, from lava flows in the maria, are still 3.1 billion years old. The dry, airless lunar surface has preserved features far older than any that survive on Earth.

timeline

4.45 Ga	4.404 Ga	4.28 Ga
Earth's crust began to solidify	Oldest mineral grain dated	Possible age of the Nuvvuagittuq greenstone rocks

Ancient surface The paler areas around the maria and across most of the far side of the Moon are the lunar highlands, the oldest rocks on the Moon and older than any on Earth. Many have been shattered and altered by later impacts, but among them remain areas of pale rock that are the remnants of the Moon's primordial crust. The Apollo 15 astronauts found a piece and called it the Genesis rock. It is a rock type called anorthosite, which probably formed as crystals grew in molten magma. It turned out to be just 4.1 billion years old – younger than expected. Samples returned by Apollo 16 give ages of 4.36 billion years, but that too is younger than expected for the oldest lunar crust. The oldest mineral grain from the Moon to have been dated is a zircon crystal 4.42 billion years old.

The first rock?

Once the snow melts in the remote tundra on the eastern shore of Hudson Bay in northern Québec, rocky outcrops are easily visible. Some of them are very ancient. Don Francis and Jonathan O'Neill from McGill University were hoping to find rocks as old as 3.8 billion years in what is known as the Nuvvuagittuq greenstone belt. But when scientists at the Carnegie Institute applied the latest dating techniques, they came back with the figure of up to 4.28 billion years! These are the oldest rocks yet identified, dating back to the Hadean period. Most of the exposure is of altered volcanic rocks, but there are also layers known as banded ironstone – rocks produced near to underwater hydrothermal vents, arguably requiring the presence of living bacteria.

4.031 Ga
Age of the Acasta gneiss

3.8 Ga
End of the late heavy bombardment and the Hadean period

> **Driven by the forces of love, the fragments of the world seek each other so that the world may come to being.**
>
> Pierre Teilhard de Chardin

Riches from heaven Although the impact craters that must have been produced on Earth by the late heavy bombardment have long since vanished, the chemical signature of that episode remains. When the metallic iron core of the Earth separated out, it took with it most of the heavy metals that are highly soluble in iron – among them gold, platinum and tungsten. Conveniently, tungsten comes in two forms or isotopes: tungsten-184 and tungsten-182. The formation of the Earth's core would have removed almost all the tungsten from the mantle; after that, the only terrestrial source would have been the decay of a radioactive element called hafnium, but that produces only tungsten-182. The oldest rocks on Earth are indeed enriched in tungsten-182. But all later rocks contain more tungsten-184. The implication is that it must have come from the sky in meteorites during the late heavy bombardment. With the tungsten would have come almost all the gold and platinum that we mine today.

Mars-sized Theia crashes into the young Earth, vaporizing a cloud of rock which forms the Moon.

The first continent?

A three-hour flight by floatplane north of Yellowknife in the Canadian Arctic takes you to a region known as Acasta. The only sign of human life there is a small shed where geologists store their tools. Above the door is a sign: 'Acasta City Hall. Founded 4.03 Ga.' Until the Nuvvuagittuq rocks were dated, this was thought to be the oldest place on Earth. The rocks here are highly altered through long and deep burial in the roots of a continent that has now eroded away.

The oldest thing on Earth Little survives on the surface of the Earth from the Hadean period, and the few rocks that do are, to use a not very technical term, fubaritic; that is, fouled up beyond all recognition! One exception to this is a mineral called zircon. Though normally found only in tiny crystals the size of grains of sand, zircons can survive repeated melting of the rocks around them, with the record of where they first formed remaining intact. The oldest zircon ever found comes from a 3 billion-year-old conglomerate of older grains and pebbles in the Jack Hills region of Western Australia. The core of that crystal is 4.4 billion years old. There are also clues from the ratio of oxygen isotopes in the crystal that it may have formed in the presence of liquid water, suggesting that at least some parts of the planet were cool enough for water to condense at that time.

the condensed idea
Heavy bombardment

04 The dating game

One of the first things people ask about a rock or fossil is 'How old is it?' Until about the middle of the last century, no one knew the answer for sure. But today there are remarkably accurate techniques for dating rocks and even figuring out the ultimate date – the age of the Earth itself.

> **❝If one is sufficiently lavish with time, everything possible happens.❞**
>
> **Herodotus**

Biblical date People have been trying to determine the age of the Earth for centuries, but the early attempts were via theology rather than science. In 1654 Bishop James Ussher, primate of all Ireland, published an estimate based on very detailed analysis of the scriptures, working back through the generations of prophets to Adam. The date he arrived at for creation was 4004 BC on 22 October at six in the evening!

Scientific guesswork By the mid-19th century, geologists and biologists realized that they needed to allow a lot more than 6,000 years for all the processes of the past. Some looked at the rate at which sediments are carried down rivers and deposited, and extrapolated from that to calculate the total depth of sedimentary rocks. Others looked at the salinity of the oceans and the rate at which salt is carried down to the sea in rivers. The eminent physicist Lord Kelvin assumed that the Earth was molten at its formation and calculated the rate at which it would have cooled. He arrived at the figure of 98 million years and, even though his estimates varied between 20 million and 400 million, that figure was widely accepted.

timeline half-lives of isotopes commonly used for dating

carbon-14	uranium-235	uranium-238	thorium-232
5,730 years	704 million years	4,469 million years	14,010 million years

ARTHUR HOLMES 1890–1965

If any one individual can be said to have won the dating game, it is Arthur Holmes. He persisted with radiometric dating after others had given up. This was before the half-life of radioactive elements was well known, before the invention of mass spectrometers and before anyone realized the importance of different isotopes. Holmes was using painstaking wet chemistry to determine the abundance of trace elements in a rock. And yet his dates for the main geological periods were remarkably accurate. In 1913 he published a booklet titled *The Age of the Earth*, an age he estimated to be 1.6 billion years. He later revised that figure using meteorites, first to 3.5 billion and then to 4.5 billion – the value that is accepted to this day.

Radioactive clock In 1902, Ernest Rutherford realized that as radioactive elements decay at a constant rate, this might be used as a clock to date rocks. Radioactivity produces alpha particles, which are the nuclei of helium atoms, so Rutherford guessed that a measurement of the helium accumulated in a rock might reveal its age. He didn't understand the finer details: for instance, he didn't realize that helium might escape from the rock. He later revised his first estimate of 40 million years to 500 million years.

It was Arthur Holmes who turned radiometric dating into a precise science, measuring the half-lives of radioactive atoms – the time it takes for half of a sample to decay – and working out the complex sequence of decays that turn uranium into lead. We now know that there are two types of uranium atom: uranium-238 and uranium-235, which decay into lead-206 and lead-207 respectively, giving two independent checks on the date.

potassium-40	rubidium-87	samarium-147
11,930 million years	48,800 million years	106,000 million years

> **Because the pathway from uranium to lead was particularly complicated, others had abandoned their researches, leaving the 21 year old research student to become the world authority on a technique that was finally to provide the planet with its authentic, scientifically determined birthday.**

Robert Muir Wood on Arthur Holmes

Weighing atoms It took Arthur Holmes several months to make his first date estimates. Today rocks can be dated in minutes, thanks to a machine called a mass spectrometer. Tiny samples are vaporized and electrons are stripped off the atoms so that they can be accelerated and deflected to different detectors, depending on their mass. It means that each isotope is weighed – or even counted – atom by atom.

Tree rings and carbon

Archaeological dates up to about 60,000 years ago can be calculated by measuring carbon-14. This isotope is made by the action of cosmic rays on carbon in the atmosphere. Once it is incorporated into living plants and animals, production stops and the carbon-14 decays with a half-life of 5,730 years. Modern instruments can measure ages up to ten times the half-life, after which too little remains to measure.

But the cosmic ray flux has not been constant. Fortunately, nature has provided a calibration chart in the form of tree rings. Each growth ring in a tree trunk corresponds to a specific year. Using overlapping sequences of rings, the record can extend back thousands of years to trees preserved in bogs. Each ring can be carbon dated. The resulting wiggling curve has recalibrated carbon dating with great accuracy.

Eternity in a grain of sand There are now many dating techniques available to archaeologists and geologists. One of them can reveal when a buried grain of sand last saw the light of day. It is called optically stimulated luminescence, or OSL. Natural radioactivity causes damage in the crystal lattice of mineral grains. That damage is healed by light, which releases energy from the crystal lattice as a glow. So, if a sample is kept in darkness until it is inside the instrument and then exposed to a brief flash of laser light, the resulting glow is a measure of how long it has been buried.

clues in a crystal

Zircon (zirconium silicate) is a popular semiprecious stone. But it is even more popular among geologists studying the ancient Earth. Its crystal lattice is such that uranium atoms are easily trapped within it, but lead is not. So the formation of the crystal from molten magma sets the radioactive clock ticking and the build-up of lead from the decay of the uranium gives a surprisingly accurate age. Better still, once formed, zircon crystals are incredibly resistant. The rocks around them can be folded, fractured, buried and even remelted, but the zircon will endure. Different zones of a zircon can give different dates from the history of the crystal. The mass spectrometers used to measure them are so sensitive that up to 100 different samples can be taken from a single zircon the size of a grain of sand.

Dating mountains Dating techniques reveal many things beyond the simple age of rocks. They have been used to track the prehistoric migration of our human ancestors. They have been used to date climate change and rising sea levels. For example, uranium dissolves in sea water and can become trapped in coral as it grows. Coral always grows in shallow water, so date the coral and you know when sea level was at about that height.

Different minerals crystallize at different temperatures, so you can work out the temperature history of a rock from the mineral grains within it. For example, zircon in Himalayan granite crystallizes at more than 800 degrees Celsius, corresponding to depths in the earth of perhaps 18 kilometres (9,843 fathoms). But muscovite mica forms at cooler temperatures and therefore shallower depths. An age difference of only 2 million years within the same granite suggests a very rapid uplift of the Himalayas around 20 million years ago.

the condensed idea
Radioactive clock

05 A tale of three planets

Our planet, the third rock from the Sun, is a 'Goldilocks' world; one that is 'just right' for life as we know it. But why should that be and why should the second and fourth rocks – Venus and Mars – have turned out to be so different? Can we learn from their mistakes?

Ugly sister Named after the goddess of love, Venus is a beautiful planet when seen from Earth, chasing the Sun as the morning or evening star. But what we are seeing are the blueish-white cloud tops at comparable pressures and temperatures to those on Earth. The reality is very different. Those clouds are made of sulphuric acid droplets and the planet's tortured surface, 50 km (31 miles) beneath them, experiences pressures 90 times greater than on Earth and temperatures high enough to melt lead.

In many ways, Venus is Earth's sister, being about the same size and density and born at the same time with the same composition. But a different upbringing has turned Venus into the evil twin. If you were to take all the accumulated limestone, chalk and coal on the Earth and vaporize it, you would end up with a carbon dioxide-rich atmosphere very similar to that on Venus. Heat causes evaporation and water vapour is a powerful greenhouse gas, trapping heat and causing more evaporation. If you were to tow our planet just that little bit nearer the Sun and let the climate stabilize, you would find that it could not do so until all the oceans had boiled away. That is very probably what happened on Venus. Today, there

timeline

1960	1972	1975	1980 and 1982
First (USSR) attempt to launch a Mars probe (failed)	Mariner 9 (USA) is the first craft to orbit Mars	Venera 9 and 10 (USSR) take first pictures of the surface of Venus	Viking 2 and 1 (USA) land on Mars

is very little water left even in its atmosphere, as sunlight has split it into hydrogen and oxygen; the hydrogen has escaped into space and the oxygen has reacted with rocks.

The big question is: could it happen here? The answer is probably not at the moment, even with the large amounts of carbon dioxide we are releasing into the atmosphere. But in another billion years or so, as the sun grows warmer, it could be a real threat to our descendants.

Geology without water Geologically, Venus looks rather similar to Earth. Granted there are no oceans or vegetation, but there are volcanoes, impact craters, mountain ranges and cracks or fault lines. However, the faults and volcanoes are spread randomly over the surface. They do not follow the lines of plate boundaries. And the impact craters are evenly spread too, suggesting that the whole of Venus's surface is about the same age. That age is about 600 million years, relatively young compared to the surface of Mars, the Moon or Mercury.

Drilling for Martians

If life still exists on Mars, it is most likely to be found beneath the surface. Perhaps there are bacteria, warmed by hydrothermal systems and living off the chemical energy of sulphide minerals. That is why in 2005 NASA scientists drilled a hole near the Rio Tinto (red river) in south-west Spain. This is no ordinary river. As its name implies, the water runs red with dissolved iron and other minerals. They are released by the activity of bacteria beneath the surface, which make the water highly acidic. Not only is this an analogue for possible life on Mars, the project also tested a remote-controlled drill that might one day be used on Mars to search underground for traces of life.

1990–94	2003	2004	2006
Magellan (USA) Orbiter maps Venus with radar	Mars Express (Europe) in orbit. Beagle 2 (UK) fails to call home	Spirit and Opportunity rovers (USA) arrive and endure	Venus Express (Europe) in orbit

The explanation may lie in the way in which Venus loses internal heat. On the Earth that process is accomplished by plate tectonics. Hot volcanoes create new crust, while old, cold crust dives back into the planet. But that process is lubricated by water. On Venus, without water, it can't happen. So the internal temperature rises to a point where volcanoes break out all over the planet, resurfacing most of it every 600 million years in spectacular eruptions.

Critical mass Mars is half the size of Earth and only twice the size of our Moon. And that could be its undoing. The surface gravity is little over a third of that on Earth, and there is no significant magnetic field to protect the top of the atmosphere from the solar wind of charged particles. So some gas molecules, notably water vapour, get split up and slowly escape into space. There have been estimates that up to 100 tonnes of Martian atmosphere is lost to space every day, even today.

Atmospheric pressure on Mars is so low that, even if it was above freezing, liquid water could only exist today in the lowest valleys. Elsewhere, ice would turn directly to vapour without melting. And without a thick carbon dioxide comfort blanket, it always is well below freezing: typically 60 degrees Celsius below.

Fossil Martians?

In 1996, a meteorite hit the headlines all over the world. Found in 1984 in Antarctica, its composition showed that it was from Mars. Tiny cracks contained carbonate, implying deposition by water about 3.6 billion years ago. But there were also chemical traces that, had they been found on Earth, would be attributed to life. There were even structures scientists claimed were fossil bacteria, though they are 100 times smaller than most terrestrial bacteria. The jury is still out on fossil life on Mars, but the hunt goes on.

The rivers of Mars Mars clearly hasn't always been so cold and dry. Space probes have mapped most of the surface in exquisite detail, revealing clear evidence of running water in the past. But most of it is probably more than 3 billion years in the past. More recent examples may be due to localized heating of buried ice by hydrothermal activity, resulting in brief flash floods.

In its youth, however, Mars seems to have had rivers, lakes and perhaps even oceans. Large areas in the northern hemisphere are at low elevation and have many features in common with the ocean floor on Earth. So where did all the water go? It is likely that much of it escaped into space, but there could also be very large quantities still present as ice beneath the surface.

Is there life on Mars? Today, Mars appears lifeless. It is certainly not inhabited by intelligent, hostile aliens. But there is a chance there could still be little green somethings on the planet. The dry valleys of Antarctica are some of the most Mars-like environments on Earth: permanently frozen and without rain or even snow for thousands of years. And yet in the pore spaces just beneath the surface of some of the stones is a thin green layer of microscopic algae. NASA decided in the 1970s that ambiguous results from their Viking landers did not show evidence of life on Mars, and there have been no successful searches there since. But it is just possible that primitive bacteria or algae are still hanging on where they can.

> **We are all … children of this universe. Not just Earth, or Mars, or this system, but the whole grand fireworks. And if we are interested in Mars at all, it is only because we wonder over our past and worry terribly about our possible future.**
>
> **Ray Bradbury,** *Mars and the Mind of Man,* **1973**

the condensed idea
Habitable zone

06 Living planet

Alien visitors approaching our solar system would know at once where to go to meet the locals. Quite apart from our radio transmissions, the signature of life is quite clear in Earth's atmosphere. Instead of a composition dominated by carbon dioxide, like an engine exhaust, there is an unstable mix including oxygen, ozone, traces of methane and ammonia – gases that could only be maintained by life.

Planetary thermostat For the past 3.5 billion years, the Earth seems to have maintained a surface temperature in a range between 10 and 30 degrees Celsius. Yet, during that time, the Sun's output has increased by somewhere between 1.5 and 3 times. We only need to look at boiling Venus or frozen Mars to see how different things might have been. That stability has been achieved thanks to the actions of bacteria and algae.

Since the origin of life, organisms have been eating the carbon dioxide blanket that kept them warm. The geological record contains repeated thick layers of limestone and chalk, essentially fossils of the early atmosphere precipitated by living organisms. Earth's atmosphere may once have been the same composition as those of Mars and Venus, with up to 95 per cent carbon dioxide. Today, CO_2 makes up just 0.03 per cent of our atmosphere.

Global pollution The first bacterial inhabitants of Earth were probably the sort of things we find today in sewers: smelly creatures deriving their energy from chemical decay and thriving in the absence

timeline

3.5 Ga	2.8 Ga	2.45 Ga	2.45–2.0 Ga	0.85 Ga
Early atmosphere rich in carbon dioxide	First oxygen released by cyanobacteria	Free oxygen starts to accumulate in atmosphere. CO_2 drops	Ice age	Oxygen levels start to rise

of oxygen. But then came one of life's great inventions: photosynthesis. Cyanobacteria, or blue-green algae, took energy from sunlight and used it to combine carbon dioxide and water into the complex chemical structures of their bodies. They also produced a waste gas, a poison toxic to their anaerobic companions: oxygen. It has been described as the worst pollution incident the world has known.

At first, from about 2.8 billion years ago, the oxygen was rapidly used up in chemical reactions in seawater. One of the products may be the extensive banded iron formations of the Precambrian, showing how the planet essentially went rusty. The layers may be due to seasonal blooms of algae boosting oxygen levels, or perhaps to upwelling currents bringing more iron in solution from the anoxic depths.

Breath of life The first evidence of free oxygen in the Earth's atmosphere comes at about 2.45 billion years ago. It was closely followed by an ice age; perhaps the algal blooms drew down a bit too much carbon dioxide from the planet's insulating blanket. After that, oxygen remained at about 3 per cent or 4 per cent of the atmosphere until about 850 million years ago, when it started to rise again. That may be what made it possible for complex animal life to evolve.

Are we Martians?

Being smaller and further from the Sun, Mars may have cooled faster and thus been habitable before the Earth. Life may have begun there. We know that meteorites can reach the Earth from Mars, and in the early solar system that must have happened quite frequently. It is possible that tiny bacteria hitched a ride and seeded the young Earth with life. In which case, we might all be exiled Martians!

0.78–0.66 Ga	0.61 Ga	0.30 Ga	Now
Ice ages	First large animals (Ediacaran)	Oxygen levels peak at up to 35 per cent	Atmospheric CO_2 at its highest level in more than 800,000 years

> 6Viewed from the distance of the moon, the astonishing thing about the earth, catching the breath, is that it is alive. The photographs show the dry, pounded surface of the moon in the foreground, dry as an old bone. Aloft, floating free beneath the moist, gleaming, membrane of bright blue sky, is the rising earth, the only exuberant thing in this part of the cosmos.9

Lewis Thomas, *The Lives of a Cell: Notes of a Biology Watcher*

For most of the last 540 million years, oxygen has made up about 21 per cent of our atmosphere; conveniently, just enough for large animals to breed but not so much that forest fires would burn out of control. One exception occurred around 300 million years ago in the late Carboniferous period, when oxygen seems to have reached 35 per cent of the atmosphere. This was the time when thick deposits of coal were laid down. It may also have enabled insects and amphibians to grow large, producing dragonflies with a 30-centimetre (12 in) wingspan.

Gaia theory Oxygen and carbon dioxide are just two of the factors that life seems to control on our planet. Gaia theory, proposed by independent scientist James Lovelock together with microbiologist Lynn Margulis, suggests that feedback mechanisms operate to maintain our planet as a hospitable place for life. Although named after the Earth goddess, Gaia theory does not suggest external or conscious control, just a series of feedback mechanisms. In addition to oxygen and carbon dioxide, factors such as methane and ammonia in the atmosphere, ocean acidity and salinity all seem to have remained remarkably constant. Life even seems to control cloud cover and perhaps rainfall by releasing dimethyl sulphide into the air, where it is oxidized to form microscopic particles which act as nuclei to seed the droplets in clouds.

JAMES LOVELOCK b. 1919

James Lovelock was born in 1919. He has never been a conventional scientist, starting out in medical research and then advising NASA on instruments to detect the composition of planetary atmospheres. Since 1964 he has worked as an independent scientist and inventor. His most famous invention, the electron capture device, was key in detecting the effects of widespread pollutants on Earth. He is the originator of Gaia theory, named after the goddess of the Earth by his friend the writer William Golding. The theory suggests that life unconsciously regulates the atmosphere and climate on Earth, a balance that humans disrupt at their peril.

Gaia the destroyer Gaia theory – that our planet behaves as a single living super organism – is still controversial, though several of its predictions have been confirmed. The goddess gives her name if not her nature to the theory, but in mythology at least, Gaia eats her own children. So what might be the effect of humanity on this self-regulating mechanism and vice versa? We are clearly making large-scale changes to our planet, clearing forests and changing land use, damaging habitats and biodiversity, releasing pollutants and unprecedented quantities of carbon dioxide. Climate models suggest that we are approaching a tipping point and that the homeostasis will re-establish around a very different, warmer world. Lovelock is even more sanguine, suggesting that Gaia will readjust conditions in such a way as to limit the human population. Sea levels might rise, vast agricultural lands may be turned to arid desert. Lovelock's rather gloomy prediction is of a world next century with a much smaller human population and the planet carrying on regardless.

the condensed idea
Self-regulating, living planet

07 Journey to the centre of the Earth

When Jules Verne published his fictional journey in 1864, the interior of the Earth was almost unknown. Geology was a new science. Charles Darwin's theory of evolution had only just been published and the first dinosaur fossils were beginning to appear in the world's museums. Today, we don't need subterranean passages for a fictional journey; with modern instruments, we can make a factual one.

Just a few miles away from where you are now is a place that no one has visited and probably never will. If the distance were horizontal, it would be just a short drive, but to travel that depth vertically presents insurmountable problems. The deepest a person has gone was on 23 January 1960, when Jacques Piccard and Lt Don Walsh reached the floor of the Challenger Deep in the Mariana trench off Guam in the Pacific, travelling in the submersible bathyscaphe Trieste. They reached the ocean floor an estimated 10,911 metres down (35,797 ft), where they saw a flatfish.

Digging to the centre of the earth The deepest mine is the Tau Tona goldmine in South Africa at 3,900 metres (12,795 ft). It is so deep inside the Earth that, without air conditioning, the temperature would be

depth

0 km	−38 km	−100 km	−670 km
Surface	Average crust thickness	Average base of lithosphere	Base of upper mantle

Life inside the Earth

Life has colonized every environment on the Earth and in the sea. But it doesn't stop there. Sediment cores drilled by research vessels from the deep ocean floor are teeming with life. The inhabitants are bacteria, mostly the primitive type known as Archaea. Their ancestors may have been buried under the seafloor millions of years ago, but they have continued to live and reproduce, albeit slowly, thriving on organic matter and methane that was buried with them or that percolates through the sediment in groundwater. They have been found more than 5 kilometres (3 miles) beneath the surface and may have been buried there for 16 million years. Altogether they make up at least 20 per cent of the biomass on Earth; some estimates put it at more than half of life on Earth.

about 60 degrees Celsius. The deepest borehole drilled is on Russia's Kola Peninsula: it reached 12,262 metres (40,230 ft) in 1989. The temperature and pressure make it impossible to go any deeper, as the semisolid rock would close up the hole as quickly as it was drilled.

There is still a possibility of drilling right through the crust into the mantle, but not on land. The ocean crust is on average a mere 7 kilometres (4½ miles) thick. In the early 1960s, there was an ambitious plan to drill right through it and reach the so-called Mohorovičić discontinuity or Moho, which marks the boundary between crust and mantle. But funding failed to materialize and the technology of the time was probably not up to the job, so the 'Mohole' was abandoned. Now there is another similar proposal, thanks to the latest state-of-the-art Japanese scientific drilling vessel, Chikyu (meaning planet Earth in Japanese). It employs what is

−2,891 km	−5,150 km	−6,371 km
Base of lower mantle	Base of molten outer core	Centre of the Earth

The Moho

In 1909, the Croatian geophysicist Andrija Mohorovičić was studying how natural earthquake waves refract through the layers of the Earth. He discovered that there was an abrupt change in the velocities of the seismic waves around 22 miles (35 km) beneath continents. Above it, compressional seismic waves (P-waves) travel at about 4 miles per second (6 or 7 km/s), below it at about 5 miles per second (8 km/s). That has become known as the Mohorovičić discontinuity, marking the boundary between the crust and mantle.

known as riser technology, in which pressurized drilling mud is pumped down a concentric outer tube. Developed for the oil industry to prevent blowouts, it could also be used to keep the hole open at great depths.

Planetary phrenology A scaled-down Earth appears to be a very smooth ball to the first approximation. If it were the size of a desk globe, even the Himalayas would only rise a fraction of a millimetre from the average surface. The same is true inside the Earth. The layers of crust, lithosphere, upper and lower mantle, outer and inner core are smooth and even, like the layers of an onion. But it is only 99.9 per cent like an onion. The 0.1 per cent of difference from those even layers is where much of the most interesting geophysics is to be found: the clues to the processes at work inside.

"Was I to believe him in earnest in his intention to penetrate to the centre of this massive globe? Had I been listening to the mad speculations of a lunatic, or to the scientific conclusions of a lofty genius? Where did truth stop? Where did error begin?"

Jules Verne, *Journey To The Centre of the Earth*

Astronauts have commented how perfectly round the Earth appears from space. But it is not exactly spherical. What do you think is the highest point on the surface, measured from the centre of the Earth – Everest? Wrong. It is a volcano in Ecuador called Chimborazo. Although it is only 6,275 metres (20,587 ft) above sea level, it is very close to the equator and the rotation of the Earth makes our planet bulge at the equator. Fortunately the atmosphere rotates with the Earth, so we don't notice the rotation speed, but if you are standing at the equator, you are heading east at just over 1,000 mph (1,670 km/h) simply due to the rotation of the Earth.

Feeling the bumps from space Launched by the European space agency in March 2009, the gravity field and steady-state Ocean Explorer (known by its initials GOCE) is measuring Earth's gravity field with unprecedented detail. In its low orbit, it is highly sensitive to the tug of gravity. Slightly more gravity and it accelerates slightly; slightly less and it slows down. The atomic clock makes it possible to measure very small differences and produce the most accurate gravity map of the Earth so far. It reveals the shape of the 'geoid' – the imaginary shape of a global ocean in the absence of tides and currents. It is a crucial reference for accurately measuring ocean circulation, sea-level change and ice dynamics – all potentially affected by climate change. An exaggerated model based on the data makes our planet looked like an irregular potato, with lumps around Indonesia and north-west Europe and a dimple in the Indian Ocean. Most noticeable is the significant dip in gravity just south of India, in the region through which that subcontinent drifted in the recent past. It means that sea level there is about 100 metres (328 ft) lower than average.

With a thin skin of crusty continents and thick layers of mantle and inner and outer core, the Earth is almost like an onion. But not quite.

As we will see in the following sections, lumps, bumps and irregularities continue down through the Earth all the way to the core, giving us new ideas about the dynamic processes going on there.

the condensed idea
Almost an onion

08 Seeing inside the Earth

We may not be able to sample the deep interior of the Earth directly, but we can certainly take a look. Our planet is opaque to light but not to the seismic waves from earthquakes. Just as light reflects and refracts as it bounces off mirrors and bends through lenses, so seismic waves bounce off the internal layers of the planet and bend through rocks of different composition.

The study of seismic waves is far more complex than simply throwing a stone into a pond and watching the ripples. For a start, there are three different sorts of wave and they travel at different speeds and therefore arrive at different times. The first to arrive are the P-waves, standing conveniently for both primary and pressure. Then there are secondary or shear S-waves, which travel at about 60 per cent of the speed of P-waves. Finally, there are surface waves, which travel at an intermediate velocity but take the long way round, around the surface of the planet. As they propagate in two dimensions rather than three, they dissipate more slowly and can circle the Earth several times. Reflected P- and S-waves interfere with surface waves, meaning that the latter can also carry information about the interior of the Earth.

Shooting at the seafloor Probing the deep interior of the Earth calls for seismic waves produced by a big bang such as an earthquake, or, in the past, an underground nuclear test. In order to study the shallower

depth

0 km	−38 km	−670 km
Surface	Average base of crust	Base of upper mantle
Temperature 20°C	500° to 900°C	2,300°C
Pressure 1 atmosphere	1,200 atm	227,000 atm

Under the crust

The Mohorovičić discontinuity marks a change in composition. As hot mantle rocks melt, only a small fraction of the material goes to produce the magma that erupts in volcanoes or is squeezed into the base of continents to provide new crust. What is left behind at the top of the mantle is a rock called peridotite, composed of dark-green olivine and other dense minerals. Above it at the base of the crust are rocks that are lighter in colour and density. In the case of ocean crust these are typically a rock called gabbro, which has a similar composition to basalt but in a coarse, more crystalline form as it has cooled more slowly.

layers in the sediments of the Earth's crust, you can make your own bang. At sea that is achieved by towing out a big compressed air gun and firing it at intervals. The reflections are picked up by towed arrays of hydrophones, and they can reveal ancient lava flows, layers of sediment and the sorts of domed structures that can trap large reserves of oil and natural gas. For shallower surveys through a few hundred metres of soft sediment, you don't need a bang at all. The acoustic ping of a sonar scan is sufficient.

Stamping on the ground On land, seismic profiling can be done using huge trucks. They are fitted with heavy metal plates, which are placed on the ground and then vibrated up and down by hydraulic rams. Apart from not blowing a crater in the ground, this has the advantage that the length and frequency of the vibrations can be controlled easily, tuning them to pick up different sorts of features at different depths. Again, the reflections are recorded on arrays of geophones. This technique has been used to build up a three-dimensional profile of the North American continent.

−2,891 km	−5,150 km	−6,371 km
Base of lower mantle 2,700° to 3,700°C 1.4 million atm	Base of molten outer core About 5,000°C 3.4 million atm	Centre of the Earth About 5,500°C 3.6 million atm

> **Before, people thought this was a ridiculous idea. I hope that I've shifted the viewpoint from ridiculous to merely unlikely.**
>
> **David Stevenson** on his plan to launch a probe to the Earth's core

Planetary body scan Every day there are dozens of small earthquakes, and nowadays there are hundreds of sensitive seismographs around the world recording their vibrations. The result is rather like a hospital body scanner in which multiple sensors record the signals from an X-ray source as it circles the body. With some complex mathematics and clever computing, the result is a 3D image of the internal organs, or in this case the interior of the Earth.

As we've seen, the Moho at the base of the crust is a strong reflector, as is the layer at the base of the rigid lithosphere, where it transitions into the hotter, softer asthenosphere, on which the slabs of the Earth's crust float.

A real journey to the core

The premise of the Hollywood film *The Core* (2003) is impossible. Pressures deep in the Earth would crush any probe that we could imagine carrying humans. But Prof. David Stevenson of Caltech has an idea, albeit unlikely, for getting instruments to the core of the Earth. It involves starting with a deep crack in the crust and pouring into it 100,000 tonnes of molten iron. He believes that the dense liquid would force its way down through the mantle over a week or two, and if it were to carry small, heat resistant instruments the size of grapefruit, they could continue to send readings back to the surface using seismic waves, all the way to the outer core where they would eventually melt.

There are weaker reflectors 410 kilometres (255 miles) and 520 kilometres (323 miles) down and a stronger one at the boundary between the upper and lower mantle. At the base of the mantle is another thin and probably discontinuous layer called the D″ (or D double prime) layer, which may be the resting place of ancient ocean crust.

Soft rock, hot rock Seismic tomography does not only reveal well-defined layers. The waves travel more slowly through soft rock than hard, and that can be a reflection of temperature. In this way it is possible to see plumes of hot, viscous mantle rock rising from deep in the mantle beneath volcanic hotspots such as Hawaii and Iceland. It also shows old, cold slabs of ocean lithosphere diving down into the mantle.

Liquid core P-waves travel through both solid and liquid, though they travel more slowly in soft material. S-waves cannot travel through liquid. Following a big earthquake, there is a shadow on the opposite side of the Earth where S-waves are not received. This shows geophysicists that there must be a molten liquid core in the Earth that, from its density, must be mostly molten iron. P-waves can penetrate that, revealing that there is a small inner core where the waves travel at higher velocity, suggesting that it is solid iron.

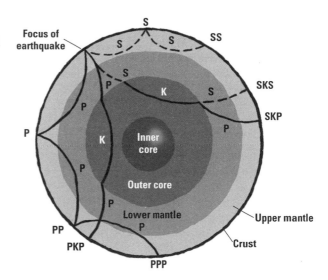

The main types of seismic wave and the different paths they take through the Earth reveal the layers of the interior.

P Primary or pressure waves
S Secondary or shear waves
K When S waves convert into P waves as they cross molten material

the condensed idea
Planetary body scan

09 Magnetic core

The Earth's core is as big as Mars but three times the mass. The outer core is a white-hot sea of molten metal with turbulent currents, eddies and storms. Beneath it is the crystal forest of the inner core. And yet processes within the core have nurtured life and also protect us from the hazardous radiation of space.

Clues from space Iron meteorites, though not the most common, are the most easily recognizable rocks from outer space. They are mostly made of iron metal but usually contain between 7 per cent and 15 per cent of nickel. The presence of nickel is often as inter-grown crystals of two alloys, one containing 5 per cent nickel, the other about 40 per cent, in proportions that make up the bulk composition. For a long time meteorites were thought of as examples of what the Earth's core might be like, having had their presumed origin in a primordial planet that broke up to form the asteroid belt. However, it seems clear that iron meteorites come from a large number of smaller bodies and that they formed at much lower pressures than those in the core of the Earth.

A better idea of the composition of the core might come from meteorites known as carbonaceous chondrites. These are the oldest objects in the solar system and the most likely to represent the bulk composition of its rocks. They contain silicate minerals in a carbon-rich matrix and also about 30 per cent to 40 per cent iron, some of it metal and some of it as oxide and sulphide. Here too, the iron is associated with nickel.

timeline

1687
Isaac Newton uses gravity to show that the Earth must have a dense core

1905
Einstein describes the origin of the magnetic field as one of the great problems facing physics

1926
Sir Harold Jeffreys uses seismic waves to show that the outer core is liquid

Measurements suggest that the density of the core is not sufficient for it to be pure metallic iron-nickel alloy. There must be 8 per cent to 12 per cent by weight of lighter elements, most probably oxygen and sulphur as they combine easily with iron.

Magnetic umbrella

Space is a hazardous place for a planet: the Earth is constantly bombarded by cosmic rays and a wind of charged particles from the Sun. The reason this does not fry our electronics or mutate our genes more than it does is thanks to the magnetic field. It forms a giant umbrella around the planet as it orbits the Sun. A few of the charged particles get trapped in the Van Allen radiation belts, while others stream in along the magnetic field lines to create the auroral displays of the northern and southern lights. But most get deflected past us and do no harm.

Heat source Convection currents in the liquid outer core, together with the much slower convection in the solid mantle of the Earth, are driven by a huge quantity of heat energy. The overall heat loss from the Earth is about 44.2 terawatts (TW), twice the total human power consumption. There are probably several sources. The biggest, accounting for about 80 per cent of the total, is the radioactive decay of elements such as potassium-40, thorium and uranium. More could come from latent heat released as the inner core freezes. Also, as pure crystals grow in the inner core, dissolved lighter elements such as silicon, sulphur and oxygen may be expelled, releasing gravitational energy as they rise to the base of the mantle.

Magnetic dynamo In 1905, Albert Einstein described the origin of the Earth's magnetic field as one of the great unsolved problems facing physicists. It was not until around 1946 that the German physicist Walter Elsasser, working in the USA, and the Cambridge geophysicist Sir Edward Bullard proposed that the field was generated by electric currents induced in the liquid outer core. In order to sustain that field, the molten iron must itself be circulating in convection currents. The Coriolis effect arising from the Earth's rotation would cause those currents to spiral and generate magnetism.

1936
Danish geophysicist Inge Lehmann uses seismic waves to show there must be a solid inner core

1946
Both Elsasser and Bullard claim the magnetic field is created by electric currents in the outer core

2010
Computer models and physical models using molten sodium simulate magnetic reversals

> **❝I can't imagine a less hospitable place for people. High pressure, white-hot temperatures – it's a nasty place.❞**
>
> **Prof Dan Lathrop,**
> **University of Maryland**

Chaotic currents To those navigating by compass, the Earth's magnetic field seems reassuringly constant, as if the world has a bar magnet at the centre. But down in the core it is far from simple. At about 3 million times atmospheric pressure and nearly 4,000 degrees Celsius, molten iron is almost as runny as water. Its convection in the outer core is complicated by local eddies somewhat similar to storm patterns in the atmosphere. Many of the wilder excesses are dampened out by the effects of the inner core and possibly by shielding due to iron caught up in the D″ layer at the base of the mantle. But some anomalies still get through.

Magnetic anomaly The south-west Atlantic has been likened to the Bermuda Triangle by space scientists. There have been many cases of instruments malfunctioning in satellites as they fly over the area. There seems to be a huge magnetic anomaly drifting slowly westwards, with half the field strength of that at the poles. As a result, charged particles from space can reach the low orbits of satellites and cause damage.

Polar reversals The South Atlantic anomaly may be the first sign of something bigger. The Earth's magnetic field strength has been decreasing over the last 180 years. It may be about to reverse entirely. Magnetized volcanic rocks show that this has happened many times in the past – on average once every 300,000 years. But it has been 800,000 years since the last full reversal and no one is sure how quickly it happens or what will happen to our magnetic protection and navigation systems while it does.

Crystal forest There is something strange going on at the centre of the Earth. Seismic waves passing through the inner core take slightly longer if they're travelling east to west, compared to those travelling north to south. The effect is even clearer in the sort of harmonic vibrations emitted when the Earth rings like a bell after a big earthquake. The best explanation for this so-called anisotropy is if the inner core is crystalline

Magnetic future

Our present understanding of the Earth's magnetic field suggests that it could not have existed without a solid inner core. So the magnetic field will be no older than that core. However, there is evidence in rocks found in Australia that they were magnetized by a field 3.5 billion years ago, so the inner core must have been forming by then. Eventually, the entire core will freeze, the magnetic field will die away and our descendants will have no magnetic protection from space radiation. But that is not likely to happen for another 3 to 4 billion years.

with the crystals aligned north–south. Experiments in a diamond anvil have shown that iron-nickel alloy crystals grow much bigger at the sort of pressures found around the inner core, and it is possible that the core is made up of interlocking crystals many kilometres long.

The orientation of that anisotropy is also slowly changing, suggesting that the inner core has rotated about a tenth of a turn faster than the planet as a whole over the last 30 years. It may be experiencing a magnetic pull from currents in the outer core analogous to the jet streams in the atmosphere.

the condensed idea
The churning core is a magnetic dynamo

10 The moving mantle

The Earth's mantle is solid rock, but it is hot and under pressure. Over geological timescales, it can flow rather like solid ice in a glacier. Heat from deep within the Earth causes convection currents similar to the churning in a pot of viscous porridge. The currents produce the forces that create earthquakes and volcanoes and drive continental drift.

The Earth has a problem with its thermostat – or at least it would if the mantle was hard and rigid. All that excess heat from radioactive decay has to get out somehow, and rock is a good insulator. Fortunately for the Earth's temperature control, mantle rocks can slowly move, transporting heat up to the crust and beyond as they do so. It's a constant battle between temperature and pressure. As rocks in the deep mantle heat up they expand, lowering their density so that, over millions of years, they begin to rise.

Double boiler One of the big problems in geophysics has been reconciling circulation in the convecting mantle with observed differences in composition and apparent layering as revealed by seismic waves. The result has been an argument between those who think that the entire mantle undergoes convection in a single circulation, and those who prefer the idea that it is a sort of double boiler with little or no exchange across the boundary between the upper and lower mantle 660 kilometres (410 miles) beneath our feet. The true situation is probably a mixture of both.

timeline the billion-year mantle convection cycle

−200 Ma	−1 ka	Now
Plume of hot rock starts to rise from the base of the mantle	120 km down, melting begins and the melt rises faster	Magma erupts along a mid-ocean ridge forming new ocean crust

Diamond anvil

The pressures and temperatures of the deep mantle are hard to imagine and difficult to simulate. The answer is to use the toughest material known to man: diamond. A diamond anvil cell is reasonably small and simple, if expensive. At its heart are two cut diamonds, which concentrate pressure into a minute sample of a mineral between tiny opposing diamond facets. Diamond has the additional advantage of transparency, so a laser can heat the sample and a microscope can view what's going on. Occasionally, a diamond will fracture, but more often the worrying 'crack' comes from the mineral sample changing phase to the higher-density form that it might adopt in the mantle.

While it's impossible to visit the upper mantle itself, small lumps of it get caught up in volcanic eruptions and bigger slabs are sometimes thrust into view by fracturing. They are made of a rock called peridotite, consisting of dense green olivine and other minerals, looking like hardened green Demerara sugar. To understand the layering of the mantle, scientists put tiny samples of peridotite into a diamond anvil cell (see above box).

Phase change Increase the pressure and temperature in a diamond anvil cell to the conditions found at the boundary between the upper and lower mantle and after a while there is a sudden 'crack'! With any luck, it is not the diamond breaking but a sudden phase change as the minerals in the peridotite adopt a new and higher-density crystal structure. The composition has not changed, but the new mineral, called perovskite, has different physical properties and, at the base of the upper mantle, shows up as a clear layer reflecting seismic waves. The minerals undergo other phase changes at about 410 kilometres (255 miles) and 520 kilometres (323 miles), depths that both correspond to reflective layers.

+140 Ma	+200 Ma	+500 Ma
The slab of ocean lithosphere, now cold, begins to sink back down into the mantle	The slab reaches 660 km and waits while mineral crystals change to a high-density phase	The dense slab breaks through into the lower mantle and falls rapidly to the core mantle boundary

> **The big controversy today is the nature of the flow of the mantle: is it just one cell that extends all the way down to the core, or is there a two layer system in which the material we see at mid-ocean ridges doesn't sink further than 700 km?**
>
> Prof. Dan McKenzie, BBC radio, 1991

Hot spots The magma that erupts from ocean volcanoes such as those along the Mid-Atlantic Ridge is very different in composition to the mantle rocks that it comes from. As a plume of warm mantle rock rises, the pressure drops and it starts to melt. But only a small fraction – typically 10 or 12 per cent – melts. It is able to percolate between grains of the mantle rock and squeezes like water from a sponge towards magma chambers near the surface. The melt has the composition of basalt and forms new ocean crust. The depleted remainder is harder and moves aside as part of the mantle lithosphere beneath.

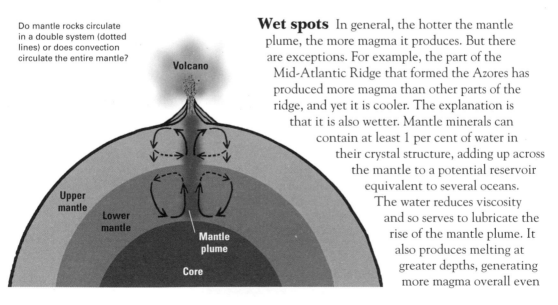

Do mantle rocks circulate in a double system (dotted lines) or does convection circulate the entire mantle?

Volcano

Upper mantle

Lower mantle

Mantle plume

Core

Wet spots In general, the hotter the mantle plume, the more magma it produces. But there are exceptions. For example, the part of the Mid-Atlantic Ridge that formed the Azores has produced more magma than other parts of the ridge, and yet it is cooler. The explanation is that it is also wetter. Mantle minerals can contain at least 1 per cent of water in their crystal structure, adding up across the mantle to a potential reservoir equivalent to several oceans. The water reduces viscosity and so serves to lubricate the rise of the mantle plume. It also produces melting at greater depths, generating more magma overall even though it is cooler. Some of the water is probably primordial, left over from the formation of the mantle, and some may be dragged down into the mantle when the ancient ocean lithosphere subducts.

Sludge at the base of the mantle Seismic reflections have revealed a thin layer at the base of the mantle that seems highly variable. In places it is up to 400 kilometres (249 miles) thick; in others it is absent altogether. They call it the D″ layer (pronounced D double prime). The boundary between the mantle and the outer core is neither clear nor even. Molten metal from the core is drawn by capillary forces into the pores between mineral grains of the mantle. Here it reacts with the silicate rocks to produce various alloys. As the mantle starts to rise, these alloys fall back under gravity to form banks of high-density sediment at the mantle base. Rich in iron, they may be conductive and generate their own magnetic fields. Tiny wobbles or nutations detected in the rotation of the Earth may be due to magnetic forces in the D″ layer.

Impossible quake

In 1994, a powerful earthquake shook Bolivia. There was little damage because it was very deep within the Earth. So deep in fact, at 640 kilometres (398 miles), that it shouldn't have happened. The rocks at that depth are just too soft to fracture. The explanation is that they didn't fracture, they shrank! The focus was in a slab of ancient Pacific lithosphere that is sinking back into the mantle beneath the Andes. When it reached that depth, the low density of peridotite meant that it could go no further. Then, in a sudden collapse, the crystal structure realigned to that of the denser perovskite, triggering the quake and allowing the slab to continue into the lower mantle.

the condensed idea
Mantle convection

11 Superplumes

The jury is still out over the extent of whole-mantle circulation. Some everyday volcanoes may have their roots in the upper mantle only. But very occasionally something happens on an altogether larger scale, rising from the core–mantle boundary, lifting entire continents and sometimes splitting them and pouring out unimaginable quantities of molten rock. This is a superplume.

Around 120 million years ago, something spectacular happened in what is now the western Pacific. Volcanoes are always dramatic in their scale and power, but none of those still active today could come close to matching what took place in the early Cretaceous. Even the great eruptions of continental flood basalt such as the one that split the Indian subcontinent and produced the Deccan Traps 65 million years ago, perhaps contributing to the demise of the dinosaurs, were trivial compared with this.

Ontong Java Plateau The Cretaceous eruptions began on the ocean floor about 125 million years ago. At their peak, they produced around 35 million cubic kilometres (8.5 million cubic miles) of basalt per million years – double the normal rate at which ocean crust is created. What they left behind is known as the Ontong Java Plateau. It covers around 2 million square kilometres (200 million square miles) of ocean floor and is up to 30 kilometres (19 miles) thick. Though now separated, the Manihiki and Hikurangi Plateaus were also once part of the outpouring. Together, they represent around 100 million cubic kilometres (24 million cubic miles) of basalt magma.

timeline superplume eruptions Approximate volume of basalt erupted (in km³)

251 Ma	200 Ma	183 Ma	138–128 Ma
The Siberian Traps (Russia): Between 1 and 4 million km³	Central Atlantic: Large eruptions as ocean begins to open	Karoo and Ferrar (South Africa/Antarctica): 2.5m km³	Paraná Traps (Brazil): 2.3m km³

The best explanation is that a superplume of hot rock rose from the base of the mantle and spread out like a giant mushroom cloud beneath the lithosphere, feeding a number of different volcanic hotspots at the same time.

Global consequences Some of the consequences of the Cretaceous superplume are still visible around the globe. The first is that there must have been a huge rise in global sea level – around 250 metres (820 ft)! That was due in part to simple displacement by the huge quantities of basalt that had erupted; the uplift of the whole region above the rising superplume may also have been a factor. This process created vast shallow seas across the more low-lying areas of continents. Unlike the deep ocean, these seas were not deep enough for water pressure to dissolve the calcite skeletons of plankton that sank from the surface. So thick deposits of chalk and limestone accumulated, giving us distinctive rock features including the white cliffs of Dover. Organic carbon also accumulated in deeper anoxic water, where it was buried and eventually matured to provide more than 50 per cent of our present-day oil reserves.

125–120 Ma	139 Ma	65 Ma	61 & 56 Ma	17–14 Ma
Ontong Java Plateau: 100m km³	Caribbean Igneous Province: 4m km³	Deccan Traps (India): 512,000 km³	North Atlantic: 2m km³	Columbia/Snake River basalt (USA): 175,000 km³

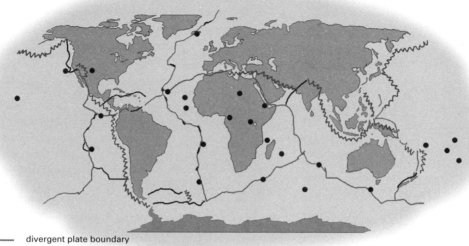

————— divergent plate boundary
————— transform plate boundary
⌇⌇⌇⌇⌇ convergent plate boundary
● hotspot

The major plate boundaries and volcanic hotspots on Earth.

Golden syrup

A popular teaching aid when describing mantle convection is a large beaker full of cold golden syrup. Localized heating at the base will cause a visible plume of hot syrup to start to rise. A cracked biscuit floated on the surface will eventually split apart as the convection reaches the surface, demonstrating continental drift. Bigger tanks of syrup in geophysics labs have been used to study the details of mantle convection.

Much of the carbon now found in Cretaceous chalk and oil probably also came from the superplume eruptions. Carbon dioxide in the atmosphere may have increased as much as tenfold, leading to a temperature rise of about 10 degrees Celsius. There is a certain irony in that by burning Cretaceous oil we may be restoring Cretaceous climatic conditions.

The next superplume There has been nothing on the scale of that superplume eruption since the Cretaceous, but could it happen again? Almost certainly. The seismic body scans of our planet reveal two plumes of hot mantle material that, at around 1,000 kilometres (621 miles) across, are potentially big enough. One is under the South Pacific, the other beneath Africa. The South Pacific plume may be the old remnant of the Cretaceous superplume; its most active days may be over.

The African superplume seems to be carrying some cold material with it and that may have stalled its progress, as may the ancient bulk of the African continent. But it may yet find a way through. An offshoot from it comes up beneath the East African Rift Valley and seems to be trying to split the continent. Perhaps one day it will produce a new ocean.

Initiating a superplume Computer simulations at Caltech have shown one way in which a superplume might arise at the base of the mantle. It involves a large slab of ancient lithosphere that has sunk right down to the core–mantle boundary. Still comparatively cold, dense and rigid, it blocks the formation of small plumes for 150 million years or more. But all the while heat is building up beneath it, and finally, perhaps after 200 million years, the superplume breaks through and rises very rapidly through the mantle over the space of just a few million years. In the simulation, it carries far more hot material than a conventional plume and forms a sort of mushroom cloud of hot, soft rock in the mantle, leading to huge eruptions on the surface over a wide area.

Another scenario for initiating a superplume also involves a descending slab of lithosphere. On its way down, it gets held up at the base of the upper mantle until the mineral grains adopt new, dense, high-pressure forms. When it finally breaks through, it collapses rapidly to the core–mantle boundary. Something has to give way to make room for it, and it displaces material that has been sitting down there heating up for a long time, which rises in a superplume.

Magnetic twist There is another twist to this tale and it extends right into the Earth's core. By removing so much from the churning outer core, the Cretaceous superplume may have calmed some of the more chaotic magnetic behaviour, leading to a 40-million-year period free from magnetic polar reversals.

> **'The field of the Geologist's inquiry is the Globe itself ... it is his study to decipher the monuments of the mighty revolutions and convulsions it has suffered.'**
>
> **William Buckland,** *Vindiciae Geologicae,* **1820**

the condensed idea
Mantle heat rising in a superplume

12 Crust and continent

The surface of our planet is covered by a relatively thin veneer of cold, hard rock: the crust. It supports us and provides us with all the raw materials for our civilization. Forming the sometimes fiery interface between Earth, air and water, the crust is also the surface expression of processes going on deep beneath our feet.

Look at the Earth from space and two very different types of surface are immediately apparent: the vast blue expanse of the oceans and the smaller but still impressive rafts of rock that make up the continents. They reflect two very different forms the Earth's crust can take. Ocean crust is typically only about 7 kilometres (4½ miles) thick and is made almost entirely of volcanic basalt, with a thin overlying carpet of sediment. Geologically, all the ocean crust is young: less than 200 million years old.

Rafts of debris Continents, by contrast, are a mess. Like the tangled blocks of used metal created by a scrapyard compactor, their rock layers are squeezed, folded, bent and twisted. This is the accumulated scum on the surface of the Earth. The heart of the continent can be very ancient indeed – up to 4 billion years old. Around it lies the accumulated debris of erosion and deposition, volcanoes and continental pile-ups.

timeline the rise of Dartmoor

310 Ma	309 Ma	200 Ma
Partial melting deep within the crust produces a granitic magma	Granite rising through surrounding rocks forms a batholith	Shrinkage and hydrothermal fluids create fissures in the granite and deposit mineral veins

Basalt

Basalt is the most abundant rock in the Earth's crust, and the same is probably true of the other rocky planets. It makes up most of the crust beneath the oceans and also underplates the continents. It is produced by the partial melting of rocks in the upper mantle, forming a mixture of around 50 per cent quartz with plagioclase feldspar and pyroxene. Traces of dark magnetite give it an almost black colour. Its fine-grained texture is the result of rapid cooling following eruptions from volcanoes, often under the sea. Gabbro is a coarse-grained rock of similar composition, sometimes found at the base of the ocean crust or injected as sheets into other rocks, resulting in slower cooling and coarser crystals.

Ultimately, ocean and continental crust are both derived from partial melting of mantle rocks. But ocean crust is darker and denser, containing more magnesium silicate and iron. When it has cooled sufficiently, it is dense enough to sink back down into the mantle. But, like cork on water, continental crust can never sink. It contains less dense silicates of elements such as aluminium and, at least in its upper half, has an average composition similar to that of granite. The lower parts of continents are less well documented, but we know that they have a bulk composition more similar to that of basalt.

The growth of continents Canada, Greenland, Australia and South Africa have ancient cores of rocks over 3 billion years old, but most continental material is younger than that. This may be due to rates of preservation rather than formation. There are several ways in which continents can grow. Mantle plumes may rise beneath them but be unable to break through, and instead end up underplating the

40 Ma	2 Ma	Now
Surrounding rocks erode away. Tropical climate encourages chemical weathering of the granite	Ice age causes physical weathering of granite into rounded blocks. Ice strips away surrounding soil	Rounded blocks of granite on tors make spectacular landscape features

continent with layers of basalt. If wet ocean crust dives down beneath a continent, the water aids partial melting, leading to a rim of volcanoes like those of the Andes and the north-west USA. The volcanic rock known as andesite is produced in the process. Recycled continental material in the form of sediments can build up around the edges of continents. But the commonest rock in the upper continental crust, making up 80 per cent of it, is granite.

The rise of granite Granite is formed through the partial melting of rocks deep within the continental crust. Melting could occur due to heat from a mantle plume, or more likely through contact with layers of hot basalt underplating the continent. Granite is rich in silica (quartz), which makes it very sticky or viscous. It used to be thought that the huge domes of granite found on every continent took millions of years to rise slowly through the surrounding rocks. But it now seems that might not have been the case. The first minerals to melt when granite forms are those that contain the most water. That lubricates the melt and makes it much runnier, so large amounts can be supplied through relatively small cracks and fissures. As a result, it is now thought that granite emplacement can happen in the geologically short timescale of thousands rather than millions of years.

Granite

The most abundant rock type in the continental crust, granite forms either by the melting of existing crustal rocks or by fractional crystallization of molten basalt. In the latter case, crystals of denser, dark, magnesium-rich minerals settle out, leaving a magma that is rich in silica. Because the intrusions, or plutons, of molten granite that are injected into surrounding rock are large, they cool slowly, producing a coarse crystalline rock with abundant quartz, along with feldspar and dark minerals such as flaky biotite mica or amphibole.

> **❝No matter how sophisticated you may be, a large granite mountain cannot be denied – it speaks in silence to the very core of your being.❞**
>
> **Ansel Adams**

The granite forces its way into the shallow crust to produce huge domed structures called batholiths. Because they are so large, the granite within them cools slowly, giving the minerals time to form the large crystals that make it a popular decorative building stone. Famous British granites come from Shap Fell in Cumbria and Dartmoor in Devon, but there are much larger formations in other parts of the world. The coastal batholith of Peru, for example, is 870 miles (1,400 km) long.

The role of water The preferential melting of wet minerals to form granite, combined with similar processes in the upper mantle to form basalt, leave the remaining rocks in continental roots dry, hard and refractory. As a result, the base of a continent extends deep into the mantle. Like a floating iceberg, there is more underneath than you can see on top; the higher the mountains, the deeper the keel.

It also follows that wherever there is water, continental rocks such as granite will form. Earth, with her oceans, has continents; dry Venus does not. A tectonically active planet would never be a world covered entirely with water because, if there was water, there would be continents to rise above it.

the condensed idea
The mighty continents: scum on the surface

13 Plate tectonics

If there is one big idea from the 20th century that transformed our understanding of the Earth, it is plate tectonics. It is not simply the idea that continents drift about on the surface of the globe, but also a complete theory of how and why they do so.

Jigsaw continents As soon as reasonably accurate world maps were available in the 18th century, people began to note the similarity in shape between the coast of West Africa and the east coast of South America; but, in the context of human timescales, rock seems so hard and the continents so vast that the idea of them once being joined together and then drifting apart seemed ridiculous. There was even a suggestion that, rather than the continents drifting, the whole earth had expanded!

Continental drift It was not until the 20th century that a few geologists, most notably Alfred Wegener, began to take seriously the possibility of continental drift. But they were still in the minority. The general understanding of the mantle at the time was that it was far too resistant to allow continents to drift through it like vast ships on the sea.

> **If the fit between South America and Africa is not genetic, surely it is a device of Satan for our frustration.**
>
> **Chester R. Longwell**

In his classic 1944 textbook *The Principles of Physical Geology*, the great British geologist Arthur Holmes even proposed a mechanism for continental drift: that the mantle, though solid, could flow on geological timescales, carrying the continents on convection currents. Others, notably Alex du Toit in South Africa, showed how geological structures clearly match up across the Atlantic, like print on two halves of a torn piece of paper. And fossils showed that the two sides were once close together. But still the idea was not widely accepted.

timeline

18th century	1858	1910	1912
First accurate maps of the Atlantic show the similar shape of continents on either side	Antonio Snider-Pellegrini maps the fit of the American and African continents	Frank Taylor in the USA suggests that continents move about the surface of the globe	Alfred Wegener in Germany proposes a theory of continental drift

ALFRED WEGENER 1880–1930

Born in Berlin, Wegener went on a two-year postgraduate expedition to Greenland, where, it is said, he witnessed sea ice breaking up, giving him the idea that continents could split apart. He was persuaded by the corresponding shapes of West Africa and South America that the two were once joined; he showed that the fit was even better if you used the continental shelf margin, rather than the present coastline. He could not suggest a mechanism for how continents could move through solid rock and his ideas were not widely accepted. He died on another expedition to Greenland.

Things began to change following the International Geophysical Year in 1957, when oceanographic surveys revealed that the mid-ocean ridges run around the world like the seam on a tennis ball (see chapter 14). At the same time, maps of the foci of the majority of earthquakes showed that they clustered in lines, sometimes along the edges of continents. These lines seemed to mark the boundaries of a collection of rigid plates covering the surface of the Earth.

Plate tectonics There are seven huge plates and some other substantial ones, together with small fragments along the more complicated joins. Not all continental margins are also plate boundaries. For example, the African plate extends west to the mid-Atlantic ridge and east into the Indian Ocean. The plates go much deeper than the crust alone, extending to include the rigid lithosphere at the top of the mantle. The mantle is typically 100 kilometres (62 miles) thick under ocean plates, though it thins to almost nothing at the mid-ocean ridges. Beneath the interiors of ancient continents, or cratons, it probably reaches a thickness of about 300 kilometres (186 miles). The base of the lithosphere and thus of the tectonic plate is not always clearly defined. It is not marked by a sudden boundary that reflects seismic waves, like the Moho, but seems to be a gradual transition from hard, brittle rock that can fracture to the softer, viscous rock of the asthenosphere below.

1927	**1944**	**1960**	**1963**	**1965**
Alexander du Toit shows how rocks of South Africa match those of South America	Arthur Holmes suggests mantle convection drives continental drift	Harry Hess suggests that new ocean floor is created along mid-ocean ridges	Fred Vine and Drum Matthews provide magnetic evidence of seafloor spreading	J. Tuzo Wilson, Jason Morgan and Dan McKenzie formulate the theory of plate tectonics

Supercontinent Trace back the motions of the plates through time and a very different world map is revealed. The southern continents come together to form a landmass known as Gondwanaland, while North America joins up with Europe and Asia as Laurasia. Separating them and opening eastwards from the Mediterranean region is an ocean known as the Tethys. Together they made up Pangaea, the supercontinent that Wegener had proposed in 1912.

Polar wandering The orientation of the continents can be traced back through time by magnetism locked within the rocks. Magnetic particles line up with the Earth's magnetic field and are then frozen into the rock when, for example, molten magma cools. As we have seen, the Earth's magnetic field sometimes reverses, but it remains more or less in line with the Earth's rotation axis, giving geologists an idea of which way was north on any particular continent at the time the rock formed. By recording the magnetic orientation of successive layers, it is possible to build up what is called a polar wandering curve – except that it is not the pole that has wandered but the continent. Sometimes, rocks on different continents come into magnetic alignment at a certain age and that can mark the time when they were joined together. In the case of Africa and South America, that was in the early Jurassic, about 190 million years ago.

The continental waltz Polar wandering curves have made it possible to trace continental movements right back into the Precambrian. Over timescales of hundreds of millions of years, the continents seem to come together and break apart in a repeating cycle that would not be out of place

J. TUZO WILSON 1908–93

After the Second World War, Wilson did important work on the crustal evolution of his native Canada, but he then became interested in evidence for continental drift from the oceans. He suggested that the Hawaiian Islands and the adjacent chain of seamounts might have formed if the Pacific floor had been moving over a fixed hotspot in the mantle. He likened the concept to lying on your back in a moving stream of water, blowing bubbles through a straw. He went on to recognize the three main types of plate margin and laid the foundations for the theory of plate tectonics.

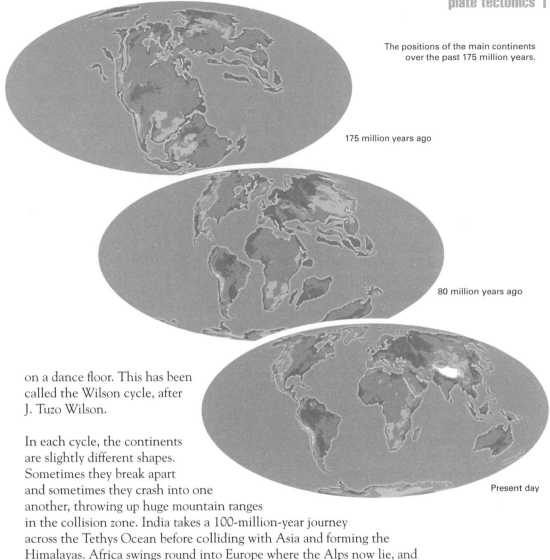

The positions of the main continents over the past 175 million years.

175 million years ago

80 million years ago

Present day

on a dance floor. This has been called the Wilson cycle, after J. Tuzo Wilson.

In each cycle, the continents are slightly different shapes. Sometimes they break apart and sometimes they crash into one another, throwing up huge mountain ranges in the collision zone. India takes a 100-million-year journey across the Tethys Ocean before colliding with Asia and forming the Himalayas. Africa swings round into Europe where the Alps now lie, and Spain hinges towards France, joining at the Pyrenees. We will return to these insights into mountain building later.

the condensed idea
Continents on the move

14 Seafloor spreading

Oceans cover more than 70 per cent of the globe's surface. We build homes and take holidays next to them; we swim, dive and fish in them; but seldom do we go deeper than the few tens of metres penetrated by sunlight. Beyond lies a whole world awaiting exploration, containing clues as to how our planet works.

The open ocean is on average more than 4 kilometres (2½ miles) deep. In the mid-19th century, ships began surveying the north Atlantic prior to laying the first transatlantic telegraph cable. Halfway across they discovered a range of mountains standing more than 2,000 metres (6,562 ft) above the abyssal plain (though still far beneath the waves). The mountains seemed to form a double ridge with a rift in between.

The longest mountain chain on Earth It was not until the 1950s, when the US and British navies wanted to survey the ocean floor for places where submarines might hide, that the full extent of the mid-ocean ridge system was realized. By this time, survey ships were fitted with sonar – a much quicker way to measure depth than lowering a weight over the side of the ship! It was discovered that the ridge system runs right the way down the middle of the Atlantic, across the Indian Ocean and through into the Pacific. Altogether it amounts to the longest mountain range on earth, 43,500 miles (70,000 km) long, snaking around the planet like the seam of a tennis ball.

timeline the Atlantic opens

180 Ma	120 Ma	63 Ma
North America begins to break away from Europe and West Africa	South Atlantic begins to open	Volcanoes in north-west Scotland as Greenland completes its separation from Europe

A future ocean ridge?

A branch of the Indian Ocean Ridge runs below the Arabian peninsula and into the Red Sea, perhaps itself a failed ocean or one yet to form. At that point the ridge branches and heads on land into Ethiopia and along the East African Rift Valley. Here it is fed by a small branch of the African superplume and there are numerous volcanoes. In the north lies the Danakil Depression, an area of Eritrea and Ethiopia that is below sea level and where the continent appears to be splitting apart. This may be the first signs of a new ocean forming.

Crustal creation The way in which the Atlantic ridge runs down the exact middle of the ocean, reflecting the shape of the coastline 1,243 miles (2,000 km) away on either side, seemed more than coincidental. In 1960, geologist and former US Navy captain Harry Hess drew the obvious conclusion: that the mid-ocean ridge was the source of new ocean crust as the Atlantic widened and the continents on either side drifted apart.

Magnetic stripes It was left to Cambridge University scientists Fred Vine and Drum Matthews to find the proof. They were able to tow a sensitive magnetometer to and fro across the mid-Atlantic ridge, mapping out the magnetic field locked in the volcanic rocks beneath. Once every few hundred thousand years, the Earth's magnetic field reverses, and they found alternating stripes of normal and reversed magnetization in the rocks on both sides of the ridge.

> **The world is the geologist's great puzzle-box; he stands before it like the child to whom the separate pieces of his puzzle remain a mystery till he detects their relation and sees where they fit, and then his fragments grow at once into a connected picture beneath his hand.**
>
> **Louis Agassiz**

56 Ma	20 Ma	Today
Magma injected beneath the ocean crust raises parts of the north Atlantic briefly to become dry land	Mantle plume at the mid-ocean ridge starts the formation of Iceland	The Atlantic is about 4,000 kilometres (2,485 miles) wide and getting wider by 3 or 4 centimetres per year

The magnetic stripes were a mirror image of each other, with the rocks getting progressively older as they moved away from the ridge itself. This was the evidence which finally convinced even sceptics of the reality of continental drift and led to the theory of plate tectonics.

Anatomy of a mid-ocean ridge Hot mantle rock is rising beneath the ridge. At depths of 100 kilometres (62 miles) or so, some of it starts to melt, rising to form the basalt of new ocean crust. The eruptions along the ridge are quite gentle, with pillow basalt oozing from fissures like thick black toothpaste from a giant tube. Eruptions take place along a rift running down the centre of the ridge. The sub-sea mountains on either side are raised up partly due to uplift from the rising hot mantle beneath and partly because they are still hot themselves. As they move aside, the rock cools, sinks and contracts, bringing the ocean floor back to its normal depth.

It is hard to have straight lines across the surface of a curved globe, which means that a number of large offsets or transform faults are found along the line of the ridge.

Black smokers There is a considerable amount of water percolating through fissures and pores in the new ocean crust. This heats up, dissolving minerals as it passes through the rocks, and can come gushing out of hydrothermal vents on the seafloor. The mineral-rich water precipitates

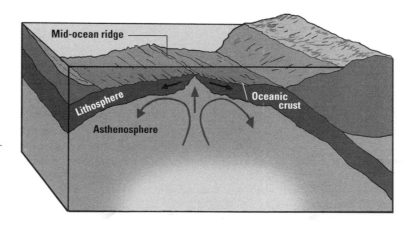

Magma rising from the mantle creates new ocean crust along a mid-ocean ridge. The new crust and underlying hard mantle lithosphere move out from the ridge on the softer asthenosphere.

challenger expedition

In 1872, HMS Challenger set sail on what was to be the world's first serious oceanographic expedition. Over the next four years, she covered 80,778 miles (130,000 km), taking deep-sea soundings, bottom dredges, ocean temperatures and cataloguing 4,700 new species. Among the sites where depth measurements were made was the Marianna trench near Guam in the Western Pacific, where a depth of 8,182 metres (5 miles) was recorded – just short of the 10,900 metres (6¾ miles) now known to be the depth of what has since been named the Challenger Deep, the deepest point in the world's oceans.

sulphide minerals as what look like clouds of black smoke in the water, building up solid chimneys around the vents. These are known as black smokers. The water can be up to 350 degrees Celsius, though at these pressures it does not boil. Bacteria and larger creatures that live off them maintain a successful if hazardous existence on the chemical energy.

The real Atlantis Iceland exists where a mantle plume coincides with the mid-Atlantic ridge. Fifty-six million years ago, as the north Atlantic was opening, there seems to have been a burst of activity in that plume. It was not hot enough to cause massive volcanic eruptions. Instead, it injected vast amounts of material beneath the lithosphere, lifting the seafloor. Seismic surveys of a 10,000-square-kilometre (2.5-million-acre) area by Cambridge scientists have revealed a lost landscape of coastlines, hills and valleys, clearly representing land that was above ground. Core samples returned pollen and lignite from forests. For one or two million years, this whole area of the north Atlantic must have been dry land. Today it lies beneath 1,000 metres (3,281 ft) of ocean and a further 2,000 metres (6,562 ft) of later sediment.

the condensed idea
Seafloor spreading from mid-ocean ridges

15 Subduction

New ocean floor is forming. Continents are drifting but not disappearing. The Earth is not getting bigger, so there must be a way of getting rid of old crust. There is, and it is called subduction. It completes the cycle of mantle circulation, along the way adding to the edges of continents and creating arcs of volcanic eruption.

There is no ocean crust older than 200 million years and very little that is older than 100 million. This is because the ocean lithosphere slab gets denser and denser as it cools and contracts until it is no longer sufficiently buoyant to float on the hot upper mantle below. So it starts to take a dive.

Ocean trenches As you follow the ocean floor across the abyssal plain away from the mid-ocean ridge, the water starts to deepen as the seafloor dives down from the relatively flat plains at 4,000 metres (13,123 ft) into a trench that can be 10,000 metres (32,808 ft) deep. There it starts to dive down out of view, but it can still be tracked by the zone of earthquakes it produces as it moves. This subduction process takes place at about the same speed that ocean crust is created along the mid-ocean ridges – between 2 and 10 centimetres (between ¾ and 4 in) per year, about the same rate at which your fingernails grow.

At first, the ocean slab dips down at an angle of around 30 degrees. Once it reaches a depth of more than 100 kilometres (62 miles) or so, heat and pressure convert the basalt into a denser rock called eclogite and the descent steepens. As the rock warms and softens, earthquakes die away, but the slab can still be tracked as a high-velocity layer for seismic waves

timeline oceans of the past

650 Ma	420 Ma	365 Ma
Panthalassic Ocean opens, with Antarctica and Australia to the north, Canada and Siberia to south	Iapetus Ocean closes between North America and northern Europe	Rheic Ocean closes between North America and South America and Florida

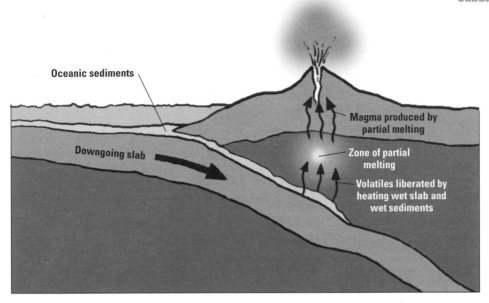

Oceanic sediments

Downgoing slab

Magma produced by partial melting

Zone of partial melting

Volatiles liberated by heating wet slab and wet sediments

As ocean lithosphere subducts, volatile components including water cause partial melting and explosive volcanism above.

from other sources. Eventually, it reaches at least a temporary halt 660 kilometres (410 miles) down, at the base of the upper mantle.

Volcanic arcs Ocean crust does not go gently. After 100 million years under the sea, it is wet – both with water in the pores and water chemically bonded in the minerals. As it descends and heats, water is driven off and rises through the rocks above, lowering their melting point. This produces a chain of volcanoes forming above the subduction zone. Where the ocean lithosphere is subducting beneath another ocean plate, it produces a volcanic island arc such as those of the west Pacific. Where it is subducting beneath a continent, the volcanoes form a mountain range on land, such as the Andes in South America and the Cascades of Oregon and Washington. Between them, these chains of volcanoes make up a ring of fire around the Pacific.

280 Ma	180 Ma	120 Ma	50 Ma
Tethys Ocean opens between India and Australia and Tibet	Atlantic Ocean starts to open	South Atlantic begins to open	Tethys Ocean closes

Future subduction Ocean crust cannot last forever. The floor of the Atlantic Ocean is still considered to be part of the European, African and North and South American plates, but eventually, the oldest parts of the ocean floor, fringing those continents, will start to subduct. The process has already begun in the western Atlantic where it meets a couple of smaller plates, forming the Puerto Rican trench in the Caribbean and the South Sandwich trench between South America and Antarctica.

Eventually, in perhaps another 150 million years, the Atlantic Ocean will begin to close again, creating a new supercontinent in perhaps 250 million years, continuing the Wilson cycle.

Not everything gets sucked back into the mantle when an ocean plate subducts. Some of the accumulated ocean sediments can get scraped off the top and accumulate at the edge of the overlying plate in what is known as an accretionary prism. This is one of the ways in which a continent can grow; it is also a means by which marine fossils can be found on land (the other being direct deposition in shallow seas when continents are flooded).

Lost oceans Great oceans of the past have subducted out of existence as continents collide. The most recent and best known is the Jurassic ocean known as the Tethys. As Africa and India moved into Europe and

HUGO BENIOFF 1899–1968

An American geophysicist working at the California Institute of Technology, Benioff was a brilliant instrument designer; his seismometers are still in widespread use today. In 1945, he used a network of ten of them in California to pinpoint the location of the first atomic-bomb test in the New Mexico desert. He went on, with Japanese seismologist Kiyoo Wadati, to plot the locations of small earthquakes beneath one of the island arcs of the West Pacific. They discovered that the earthquake zone dipped down at an angle of about 30 degrees beneath the islands. We now know that this Wadati–Benioff zone tracks the descent of a subducting slab of ocean floor.

Asia, the floor of the Tethys subducted down to the north. Seismic tomography has revealed fragments continuing their descent into the mantle. Sediment scraped from the floor of the Tethys now forms parts of the foothills of the Alps and Himalayas. Part of the mid-ocean ridge of the Tethys is to be found on Cyprus, where the rich copper ores deposited along the ridge helped to provide raw materials for the Bronze Age.

Regions where continents have collided following the closure of ancient oceans are known as suture zones. The Southern Uplands of Scotland mark one such zone, the site of the Iapetus Ocean, which closed 420 million years ago. Today the area can be crossed in a short drive from England into Scotland; 500 million years ago it would have represented a long sea voyage and parts of what are today Scotland would have been American. Caught between the two continents was the ocean island of Avalonia, parts of which are now to be found in south-west England, Newfoundland and New England.

Fossilized Moho Old, cold ocean crust normally sinks back down into the mantle in subduction zones. But in a few rare cases, sections get lifted up and eroded so that geologists can examine fossilized sections of the Moho. These are known as ophiolites, and a fine example exists in Oman, on the eastern edge of Arabia. It represents a section of ocean floor from the Cretaceous period and reveals pale-coloured gabbro overlying dark, dense peridotite. But it is not a clean boundary. Over several metres, layers of the two rock types intermesh in veins a few centimetres thick. Modern seismic reflection techniques reveal that the present-day Moho has similar complexities.

> **We are tied to the ocean. And when we go back to the sea, whether it is to sail or to watch – we are going back from whence we came.**
>
> **John F. Kennedy**

the condensed idea
Subduction and where oceans go to die

16 Volcanoes

Volcanoes are among the most spectacular surface expressions of the heat and energy in the Earth's interior. Once thought to be the chimneys of hell or the lairs of dragons, they can range from spectacular tourist attractions to deadly explosions with the potential to destroy cities, disrupt regions and change climates. They can be studied, understood and even predicted, but never prevented.

We've heard already about what lies beneath: the processes in the mantle that can give rise to volcanoes. These include the gentle upwelling beneath mid-ocean ridges that creates new basaltic crust; the huge outpourings of lava above mantle plumes; and the more explosive eruptions of wet magma above subducting plates. Now let's look at the anatomy of these volcanoes in more detail.

Volcanic entertainment on Hawaii The Hawaiian islands sit on top of a mantle plume. The Pacific plate is moving across the plume, so there is a chain of islands and underwater seamounts stretching away to the north-west, plotting its activity over the last 80 million years, with a pronounced change of direction from a more northerly drift about 48 million years ago.

Measured from its true base, 5,000 metres (16,404 ft) below the sea, the big island of Hawaii is the largest mountain on Earth. There are two main volcanic peaks: Mauna Kea, which is home to an international observatory and is hopefully extinct, and Mauna Loa, which last erupted in 1984.

timeline some famous eruptions

2.1 Ma	68,000 BC	1627 BC	AD 79	1707	1815
Yellowstone eruption: Creates 2,500 times more ash than Mt St Helens in 1980	Toba, Indonesia: Threatens early human population	Santorini, Greece: Contributed to the end of the Minoan civilization	Vesuvius, Italy: Destroyed Pompeii	Mount Fuji, Japan	Tambora, Indonesia: Volcanic dust leads to 'year without a summer'

Volcanic rocks

The nature of volcanic rocks is governed by their composition, how they are erupted, how much gas they contain and how quickly they cool. Rapid quenching creates a volcanic glass called obsidian. Lava frothing with gas that is cooled before the gas escapes creates pumice – so light that it can float. Pulverized rock can fall as ash or cinders, but if it is still soft when it lands, the fragments can weld together to form a welded tuff or an ignimbrite. Bigger lumps of lava that cool as they fly through the air form bread-crust bombs; those that are still soft when they land look more like cow pats. The surfaces of lava flows can be covered with angular blocks that are definitely not pedestrian friendly. Where there is only thin skin on flowing lava it develops a rope-like texture.

On its flanks is the crater of Kilauea, which has been in almost continuous eruption since 1991. Offshore to the south-east is Loihi, the youngest Hawaiian volcano, which has not yet broken the ocean surface but may become the next Hawaiian island.

The rising mantle rocks begin to melt deep beneath the surface, around 150 kilometres (93 miles) down. At that depth, the pressure is such that only about 3 or 4 per cent of the rock melts, producing a very runny basalt lava typical of what are known as shield volcanoes, with long, wide lava flows. Because it is so runny, escaping gas causes the lava to fountain rather than explode, so that eruptions are normally safe enough for tourists to watch from viewing platforms at the edge of the crater. The lava flows can travel considerable distances and sometimes block roads on their way to the sea.

Champagne volcanoes Altogether more violent in nature are the volcanoes above subduction zones. The magma that feeds them is produced at shallower depths from wetter rocks more rich in silica. That makes it much more viscous, so it can't flow or fountain in the same way as Hawaiian

1883	**1902**	**1943**	**1963**	**1980**	**1991**
Krakatau, Sumatra: Destroys entire island and triggers tsunami	Mount Pelée, Martinique: Town of St Pierre destroyed by pyroclastic flow	Paricutin, Mexico: Completely new volcano appears in a farmer's field	Surtsey, Iceland: New volcanic island begins to form	Mount St Helens, Washington State: Explosive eruption	Mount Pinatubo, Philippines: Aerosol cloud cools climate

Anatomy of a volcano

Mount Etna in Sicily is a complex volcano and one of the most studied. It is only 250,000 years old and already 3,330 metres (10,925 ft) tall and still growing. In fact, it seems to have been erupting more and getting more explosive in the last 50 years. It is fed by a mantle plume but not through a simple single vent. There are multiple craters around the summit and frequent eruptions through fissures around the sides. Careful monitoring of how the mountain swells and subsides and how gravity rises and falls on its flanks makes it possible to monitor the magma rising within it. Etna is so massive that there is a risk of collapse on the flanks, leading to a more catastrophic eruption.

volcanoes. The water and other volatiles make it much gassier. As magma rises and the pressure drops, it is like uncorking a shaken-up bottle of champagne full of gas bubbles. But the magma is too sticky for the gas to escape, resulting in explosive eruptions that can hurl ash and cinders high into the atmosphere. This was the sort of eruption witnessed by Pliny the Younger in AD 79 on Vesuvius, an eruption that killed his uncle and destroyed the towns of Pompeii and Herculaneum.

Subduction zone volcanoes are often stratovolcanoes: classic conical peaks like Mount Fuji in Japan, built up of alternating layers of ash and lava. Sometimes the ash cloud is lifted 20 kilometres (12½ miles) or more into the stratosphere and rains down across a continent. Sometimes the hot ash burden is greater and it hugs the ground, but, buoyed up by hot expanding gas and steam, it behaves more like a liquid, racing down the slopes at hundreds of kilometres per hour, engulfing all in its path. Such a deadly pyroclastic flow, or *nuée ardente*, descended on the town of St Pierre in Martinique on 8 May 1902, killing 30,000 people. Among the few survivors was a prisoner confined to an airless cell.

Taking the lid off On 18 May 1980, Mount St Helens in Washington State suffered the most violent eruption in recent American history. For two months the volcano had been spouting ash and steam and the north side of the mountain had begun to bulge alarmingly. When the bulge collapsed in an avalanche, it exposed magma in the heart of the volcano, instantly releasing the pressure and resulting in a catastrophic blast travelling at more than 621 mph (1,000 km/h) and toppling trees up to 19 miles (30 km) away. 1.4 cubic kilometres (0.3 cubic miles) of rock was pulverized, leading to an ash fall of up to 10 centimetres (4 in) over most of the north-west USA.

Volcanoes and climate Volcanic gas and ash clouds can have a significant effect on the Earth's climate. The unpronounceable Icelandic volcano Eyjafjallajökull caused temporary disruption to air traffic over north-west Europe in 2010 with a cloud of abrasive ash that was potentially damaging to jet engines. The ash cloud was created because the volcano erupted beneath a glacier, causing a violent reaction with melt water. But the effects did not last long.

> **❝I look back: a dense cloud looms behind us, following us like a flood poured across the land … The fire itself actually stopped some distance away, but darkness and ashes came again, a great weight of them.❞**
>
> **Pliny the Younger,** AD 79

The 1991 eruption of Mount Pinatubo in the Philippines threw about 10 cubic kilometres (2.4 cubic miles) of ash into the sky, along with 20 million tons of sulphur dioxide. The ash reached the stratosphere and spread around the planet, producing a fine haze of sulphuric-acid aerosol that was responsible for a drop of half a degree in global temperatures over the next two years, together with depletion of the ozone layer.

It is likely that bigger eruptions in the past had even greater effects. There seems to have been a significant reduction in the population of our newly evolved ancestors 70,000 years ago that coincided with a huge eruption of the Toba supervolcano in Indonesia. The vast eruptions that produced the Deccan Traps in India 65 million years ago and the Siberian Traps 250 million years ago may have contributed to mass extinctions at those times.

the condensed idea
The power of molten magma

17 Earthquakes

Continents drift around the globe with unstoppable momentum, but their edges are not well lubricated. Sometimes they get stuck. Sometimes they may slip suddenly and the ground shakes with an earthquake. When and where a continent will slip next and with what consequences are questions geologists are working hard to answer.

The relative motions of the tectonic plates that make up the Earth's crust can be tracked with millimetre accuracy using the latest techniques of global positioning, laser ranging or radio astronomy. But that's only the middle of the plates. Towards the edges, things get confused. Plate boundaries are seldom neat, straight lines: the cracks or faults get offset by other faults; multiple faults can run parallel to each other; others can branch and divide. They can stick for decades or centuries and then suddenly release the pent-up stresses in a devastating earthquake in which the ground can move tens of metres in seconds.

The San Andreas Fault Perhaps the most famous crack in the world is the San Andreas Fault in California. It is in fact a complex network of faults running through the middle of San Francisco and southwards through the hills behind Los Angeles. The Pacific plate is moving relentlessly north relative to the North American plate. In another 20 million years, Los Angeles will be alongside San Francisco.

In 1906, San Francisco was hit by a devastating earthquake that, together with the fires that followed, all but destroyed the city. There have been many

timeline major earthquakes, magnitudes and casualties

AD 526	1556	1730	1737	1755	1906	1908
Antioch, Turkey (possibly 250,000 deaths)	Shensi, China (830,000 deaths)	Hokkaido, Japan (137,000 deaths)	Calcutta, India (300,000 deaths)	Lisbon, Portugal (up to 100,000 deaths, many due to tsunami)	San Francisco magnitude 7.8 (over 3,000 deaths)	Messina, Italy magnitude 7.1 (123,000 deaths)

Magnitude

Earthquakes can occur at any depth down to several hundred kilometres. The centre of the actual fracture is called the focus or hypocentre; the point on the Earth's surface directly above that is the epicentre. The severity of earthquakes is today measured as something called the moment magnitude (rather than the old Richter scale, although the two roughly correspond). It is a function of the amount of slip, the area that slipped and the rigidity of the rocks. It represents the energy released at the focus, so damage at the surface also depends on depth. It is a logarithmic scale, meaning that a quake that is two points higher on the scale is 1,000 times more powerful.

smaller quakes since, some of them severe, such as the Loma Prieta quake near San Francisco in 1989. But California is still waiting for the 'big one'.

Where the Pacific takes a dive Japan, too, is waiting. In Japan most quakes are caused by the Pacific plate diving down beneath their east coast, giving them more than their fair share of shaking. Tokyo was hit in 1923, resulting in 143,000 deaths. The Kobe quake in 1995 killed just over 6,000, but the Tohoku quake and the devastating tsunami that followed it in March 2011 was one of the most powerful quakes in modern times. It took more than 20,000 lives and caused devastating damage probably costing hundreds of billions of dollars.

Predicting the inevitable It is easy to predict where earthquakes will occur: just look at a map of the plate boundaries. Much harder is to say when they will happen. Instruments can sometimes tell you where

1923	1960	1964	2004	2010	2011
Kanto, Japan magnitude 7.9 (142,000 deaths)	Valdivia, Chile magnitude 9.5 – most powerful recorded (3,000–5,000 deaths)	Prince William Sound, Alaska magnitude 9.2 (131 deaths)	Indian Ocean, Indonesia magnitude 9.1 (230,000 deaths)	Haiti magnitude 7.0 (220,000 casualties)	Tohoku, Japan magnitude 9.0 (over 20,000 deaths)

> ## ❛We learn geology the morning after the earthquake.❜
>
> ### Ralph Waldo Emerson

the stress is building up. Historical records reveal which segments of a fault haven't moved for a long time. With luck, seismologists can say if a major quake is likely within, say, a decade. But even if they are certain of that, it's still only a one in 3,650 chance that it will happen tomorrow – probably not a reason to engender panic or call for an evacuation.

Building for earthquakes But there are ways to prepare. It is sometimes said that it is buildings that kill people, not earthquakes, and both Japan and California now have strict building codes to minimize the risk of a catastrophic collapse. Some other earthquake-prone regions across Asia, South America and even parts of Europe are less well prepared, as shown by statistics such as those of the Armenian earthquake in 1988, which killed more than 100,000 people; by contrast, in the Loma Prieta quake in California of about the same magnitude a year later, only 62 died.

A particular hazard in some regions with alluvial soil is liquefaction. If a quake shakes up wet sediments, they can turn into something like quicksand, no longer supporting the roads and buildings on top of them. They can even amplify the earthquake waves, as happened in Mexico City in 1985. In cities, liquefaction and the quake itself can easily fracture gas and water mains, simultaneously fuelling any fires and removing the means of extinguishing them. San Francisco now has intelligent pipes that automatically shut sections off if there is a big pressure drop.

Early warning There is seldom enough certainty to issue evacuation warnings, but an increase in minor quakes can represent the foreshocks of a big one and may be a good enough reason to shut down potentially hazardous oil, chemical and nuclear plants and move emergency vehicles clear of buildings. Sometimes a brief warning is even possible after the quake has started. If the epicentre is distant from a city, radio waves can

Strange precursors

Attempts to predict earthquakes have relied on all sorts of precursors, from bizarre folklore to sensible science. Unusual animal behaviour has been reported just before an earthquake, as have sudden changes of the water level in wells. Such clues were used to evacuate the Chinese city of Haicheng in 1975 hours before a devastating earthquake. But a year later, 240,000 people died without warning in Tengshan. Scientists have monitored levels of radon gas squeezed out of rocks and looked for tiny flashes of Piezo-electric (an electric charge produced when a crystal is squeezed; as used in many gas lighters) light as mineral crystals are squeezed; while long-wavelength radio waves have also been said to precede a big quake. But none of these appears to be a reliable indicator. When a fault is ready to break, a high tide or heavy rainfall may be enough to make the difference, but who is to say when it is ready?

cross the distance a few minutes quicker than the earthquake waves themselves. In the case of a tsunami, warnings of an hour or more might be possible. One of the reasons for the terrible casualties caused by the Boxing Day tsunami in 2004 in the Indian Ocean was that the affected countries didn't have the early-warning system that was already in place around the Pacific.

the condensed idea
Earthquakes: inevitable but unpredictable

18 Mountain building

Sometimes continents collide head-on. And a slab of rock the size of a continent doesn't stop moving easily. In fact, it hardly slows down at first. When an irresistible force hits an object as immovable as a continent, something has to give. The results of these intercontinental traffic accidents are mountain ranges, and the crumple zone can extend for hundreds of kilometres.

There are three main types of convergent plate boundary: ocean subducts under the ocean creating an island arc, such as the Aleutians; ocean subducts under a continent creating a partly volcanic mountain range, such as the Andes; and continent meets continent creating the biggest mountains of all, such as the Himalayas.

The Andes The Andes are a textbook example of mountain building due to subduction. Not only does this produce a chain of volcanoes spouting silica-rich andesite, it also results in the formation of large amounts of granite, which intrude into the crust and uplift it. The mountain building phase began in the Cretaceous period around 100 million years ago and continues with earthquakes and volcanoes to this day. Across the Drake Passage, the Andes continue in the mountains of the Antarctic Peninsula. To the east of the mountains the crust has lowered, producing sedimentary basins. This may in part be due to the downward pull of the subducting slab of Pacific Ocean crust.

timeline major mountain building periods

490–390 Ma	350–300 Ma	370–280 Ma
Caledonian mountain building as North America collides with Europe	Appalachian mountain building as Africa and North America meet	Hercynian or Variscan mountain building in Europe and North America as continents collide

Further north, in the western USA, the situation is more complex. The subducting ocean crust is not descending so steeply, while the mountain ranges extend further inland and include a more low-lying, stretched area roughly corresponding to the state of Nevada, where the crust has extended.

Intercontinental traffic accident Eighty million years ago, India broke away from the southern continent and headed north. The crust of the intervening Tethys Ocean began subducting underneath Asia. Thirty million years ago, the continents met head-on and, like a car crash in slow motion, the collision is still continuing.

Ocean crust is dense enough to subduct; continental crust is not. It is like trying to hold a cork underwater. Or, to use our traffic analogy, the lower

The African superswell

It is not only horizontal collisions that can uplift the landscape. Southern Africa hasn't been in a continental collision for 400 million years and yet has been steadily rising for the last 300 million: it is currently about a mile higher than it should be were it simply floating on the mantle. The answer lies in the mantle superplume, mentioned before, which rises beneath Africa. It is a continent that has been sitting on the heat for 300 million years, pushed upwards by rising mantle rock beneath. Similarly, horizontal, unfolded layers of marine sediment in the middle of continents away from mountain ranges must have got there when the whole region was pulled down below sea level behind an ancient subducting plate. On geological timescales, continents bob up and down like corks.

100 Ma–present
Andean mountain building as Pacific subducts under South America

45 Ma–present
Alpine and Himalayan mountain building

car in an accident cannot dive down into the road. Instead, the overlying plate rises up through buoyancy, lifting not only the crumple zone of the mountain ranges but also the Tibetan plateau beyond.

When you look at the join or suture between India and the rest of Asia, it looks as if there was a convenient India-shaped indentation for the subcontinent to slot into. Not so. Simulating the collision by sliding a rigid block into a thick sheet of wet clay creates a network of criss-crossing cracks ahead of the block, along which slabs of clay are squeezed out to the side in a process called tectonic extrusion. That is what happened to Asia, pushing Indo-China out to the east.

Rapid uplift Just how quickly the Himalayas have risen can be seen from the minerals within them. Plutonic rocks such as granite cool quickly when they rise in the crust, so if their temperature can be taken at different times in the past you can date their elevation. Fortunately, that is possible. We have already heard how zircon crystals trap uranium when they form, starting a radioactive clock as the atoms decay into lead. Zircon crystallizes at more than 1,000 degrees Celsius and that can equate to a depth of 18 kilometres (11 miles). Other minerals have different so-called closure temperatures: 530 degrees Celsius for hornblende, 400 degrees for rutile, 280 degrees for biotite mica, for example. When a uranium atom does decay, it causes microscopic damage to the crystal containing it, but that damage heals or anneals above a certain temperature. For apatite that can be as low as 70 degrees Celsius. For zircon it is around 240 degrees. So the rocks carry both clocks and thermometers.

> **It must have appeared almost as improbable ... that the laws of earthquakes should one day throw light on the origin of mountains, as it must ... that the fall of an apple should assist in explaining the motions of the moon.**
>
> **Sir Charles Lyell,**
> *Principles of Geology*, **1830–3**

The story they tell in the Himalayas is one of very rapid uplift. Mount Everest and its neighbours sit on a wide bench of granite. Around 20 million years ago, that seems to have risen more than 20 kilometres (12½ miles) in little over a million years. That is an astonishing rate of 2 centimetres (1 in) a year. It may be more than the collision alone could achieve;

it could be the result of Tibet 'bobbing up' after cold, dense rocks at the base of the lithosphere broke away into the mantle. In some parts of the Himalayas, uplift continues today. The Nanga Parbat massif in Pakistan is still rising at about a centimetre per year. Sediments in the Indian Ocean suggest that the South Asian monsoon also began around 20 million years ago as atmospheric circulation was affected by the new mountain range.

Alpine uplift As the western extension of the Tethys Ocean closed, and at about the same time as the Himalayas were rising, Italy was crashing into Europe and forming the Alps. Though still spectacular, they form a smaller mountain range that is easier to study. To the north and south, thick wedges of sediment have accumulated. In between, the overlying sedimentary rocks have been spectacularly folded and contorted like so much whipped cream. So intense are these so-called nappe folds that they have sagged to the north like giant tongues, bringing older rocks into a position above younger sediments, contrary to the normal rules of stratigraphy. Where there has been most uplift, the sediments have eroded away to reveal a crystalline basement of granite and metamorphic rocks.

A leaf from Tibet

Plant leaves can be surprisingly good indicators of climate. Almost regardless of species, plants in desert environments have small, narrow leaves, while those from wet tropical rainforest have big leaves. The degree of serration of leaves to allow heavy rain to run off without damage to the leaf is another indicator. Fossil leaves from South Central Tibet indicate that the region was uplifted to its present high, dry altitude around 15 million years ago.

the condensed idea
Intercontinental crumple zone

19 Metamorphism

Rocks can be erupted or intruded into the crust. They can get eroded, dissolved and redeposited on land or under the sea, and they can be uplifted into mountain ranges. But that's just the start of their troubles. Sooner or later they are going to get buried and compressed, cooked and contorted underground. The results are called metamorphic rocks, and they may carry few traces of the original rock fabric.

To those trying to trace the remains of the earliest continents or even trying to find the remains of early life within them, metamorphism is a pain and an impediment. Some scientists refer to such rocks as fubaritic – fouled up beyond all recognition! But to a metamorphic petrologist, those textures tell the detailed history of the rock since its formation.

Heat without pressure The simplest way in which rocks can be altered by heat is known as contact metamorphism. This takes place at shallow depths in the aureole around an intrusion of igneous rock, such as granite, and is simply due to the surrounding rocks, normally sedimentary, being cooked by the heat of the cooling rock.

One of the consequences of that is that water will often be driven off, either just the water contained within the sediment, or water that is chemically bonded into minerals such as clays. Or water may be drawn into a hot igneous rock. That can, in turn, lead to another sort of local change: hydrothermal metamorphism. Typically that does not need

metamorphic zones temperature and pressure

120°C	225°C	300–900°C	150–400°C	330–550°C
Low pressure: some diagenesis consolidates sediments	Low pressure: zeolite	Low pressure: hornfels	High pressure: blueschist	Medium pressure: greenschist

especially high temperatures: between 70 and 350 degrees Celsius will do. The mineral grains in the rock may not be much affected, but new material can come in to cement them together or produce mineral veins within them. This is how the world's largest copper deposits were produced around granitic intrusions in the Andes; it is also how the china clay deposits of Cornwall came into being.

Pressure without heat Another change that can happen very suddenly to rocks is impact metamorphism. That is more commonly found on the Moon than on Earth: much of the lunar highlands have been changed by impacts. The result can be localized melting into glassy fragments or even vaporization. More common is shock deformation: shattering the rocks and perhaps welding them together again. Something similar can occur when rocks are subjected to extreme shear stress – in fault zones, for example – where it is known as dynamic metamorphism. Here, directed pressure is high but temperature is low.

Regional metamorphism By far the greatest bulk of metamorphic rocks arise from regional metamorphism, where an entire sequence of rocks has become buried deep within the crust and altered by both heat and pressure in varying degrees. These processes lead to suites of rocks described as metamorphic facies: rocks that can have had any original composition but have all been subjected to similar conditions of heat and pressure.

PENTTI ESKOLA 1883–1964

Finnish geologist Pentti Eskola invented the concept of metamorphic facies – the idea of identifying metamorphic rocks according to the conditions under which they were created, regardless of what the original rocks were before metamorphosis. His classic work came in 1920 after a year spent in Norway during which he compared metamorphic rocks there with those in his home region of Finland. His work was so highly acclaimed that he was given a state funeral.

550–700°C	600°C	300–800°C	700–900°C
Medium pressure: amphibolite	Medium pressure: melting point of wet granite	High pressure: eclogite	Medium to high pressure: granulite

Geological cookery Thanks to recent sophisticated experiments in pressure vessels and furnaces, together with theoretical calculations, petrologists now have a pretty good idea of the changes that happen at specific temperatures and pressures and can therefore work out what hellish torments metamorphic rocks have been through.

Cooks will have an idea of the sort of processes involved. If you are making a Christmas pudding, for example, and put the mixed ingredients into a pressure cooker, grains of suet will melt and, together with the flour, produce a cement-like matrix around the fruit and peel. Confectioners may start with sugar crystals, which they melt to combine with other ingredients in new crystalline forms.

Grades of metamorphism The first rocks to change as heat and pressure mount are the sedimentary clays, shales and mudstones. These are mostly clay minerals, which contain a lot of water and are easily changed by heat and pressure, so the minerals in the resulting pelite

Marble

Geologically, marble is metamorphosed limestone or dolomite. The calcium or magnesium carbonate in marble has recrystallized, and little of the original sedimentary structure remains. Pure marble is white, but it can often be coloured by veins richer in elements such as iron. The term marble is sometimes used more broadly in an architectural context to describe a wide variety of decorative building stones suitable for sculpture. The most famous white marble comes from Carrara, in the Italian region of Tuscany. This stone was highly prized for sculpture in the classical world and was the favoured medium of the Renaissance sculptor Michelangelo, who used it for his famous statue of David. Carrara marble can also be found in Trajan's Column in Rome, Marble Arch in London and Harvard Medical School in the USA.

> **'There is nothing casual in the formation of Metamorphic Rocks. All strata, once buried deep enough, (and due time allowed!) must assume that state,—none can escape. All records of former worlds must ultimately perish.'**

Sir John Herschel

are good indicators of the grade of metamorphism. The first mineral to appear in low-grade metamorphism is chlorite. As the heat and pressure mount, it gives way to biotite mica and then to garnet and so on. During metamorphism, shale can become slate and limestone can become marble. Although the silica in sandstone is chemically very stable at high temperatures and pressures, it can begin to recrystallize, binding the grains together to form quartzite.

Metamorphic texture Metamorphic rocks can develop very different textures from their parent rocks. Pressure can make thin, flat grains of, for example, mica align in a direction perpendicular to pressure, producing, in this case, a mica schist. Even fine-grained rocks such as shale can develop metamorphic texture, in this case as slate. So complete is the alteration of the original sedimentary texture of the shale, that the sheets into which slate will split, known as the cleavage, are often at a completely different angle from the original layers of deposition.

As the temperature and pressure mount, so crystals can become elongated, producing a linear texture. And they can begin to melt and recrystallize, making it remarkably difficult to tell even if the original rock was igneous or sedimentary.

the condensed idea
Changed by heat and pressure

20 Black gold

With the exception of sunshine, all the energy we use comes from the Earth. The geothermal energy from hot springs and boreholes, all our nuclear fuel, and the coal, oil and gas we burn in our homes, our power stations and our cars, and on which civilization depends – all of these have geological origins.

For more than a billion years, life has proliferated on Earth, soaking up sunshine and using the energy to build complex hydrocarbons. Many are eaten and recycled by other organisms, but eventually their remains get buried and can slowly transform into fossil fuels.

Fossil trees Three hundred million years ago in the Carboniferous period, large areas of the land were covered by forested swamps. Giant tree ferns and cycads grew and flourished, died and decomposed. Their remains built up thick layers of peat, which were subsequently buried and compressed, turning eventually to coal. This period takes its name from the vast quantities of carbon deposited both as coal and as calcium carbonate in limestone.

Fossils under the sea It takes special conditions to make oil. Fortunately for us, those conditions have been quite common in the past. The first requirement is a shallow sea teeming with life. As the micro-organisms die, with luck, they sink down into a zone with little oxygen to aid their decomposition. Ideally this will be a sedimentary basin such as the North Sea, where the crust is being gently stretched, causing it to sag down and accumulate more and more sediment. Two things happen as a

timeline short history of fossil fuels

*c.*300 BC	1775	1825	1908	1920
First recorded use of coal for metal smelting by ancient Greeks	James Watt patents his improved steam engine; underground coal mining proliferates	Commercial oil production begins in Russia	Henry Ford makes the first mass-produced motorcar	USA overtakes Russia as biggest oil producer

result: firstly, the organic remains become buried deeper and deeper and thus are subjected to more and more pressure; and secondly, the stretched crust is heated from below, so the remains are cooked. Living bacteria in the buried sediment probably also play an important part in the maturing of oil and natural gas.

This produces oil and gas, but there is one more requirement. The hydrocarbons are low density and tend to rise through porous rocks, so they need to be trapped in order to accumulate. Fortunately, layers of clay can achieve this, as can salt, which has the added advantage of itself tending to rise through sedimentary layers, forming a dome under which oil and gas can get trapped – as it has happened in the Gulf of Mexico.

Extracting oil The oil industry has made huge contributions to our understanding of geology, especially in comparatively shallow continental shelf regions offshore. There are seismic surveying techniques from ships that can penetrate the sedimentary layers with sound waves to considerable depths, revealing the layers and structures within. Oil

Storing carbon

As politicians struggle to reach agreement on reducing carbon dioxide emissions from burning fossil fuels in order to limit climate change, geologists are exploring ways of disposing of carbon dioxide from power stations. It is possible that it could be held frozen by the high pressures of deep ocean trenches. Another solution being explored off the coast of Norway is pumping it down old oil and gas wells. It could, potentially, replace the fossil fuels that produced it and even help to recover the last traces of oil and gas.

1967	1984	1988	2008
First North Sea gas field comes online	Miners' strike hits UK coal production. Pit closures follow	Intergovernmental Panel on Climate Change warns of effects of burning fossil fuels	Cost of crude oil passes $100 per barrel for the first time (a tenfold increase in a decade)

> **We've embarked on the beginning of the last days of the age of oil. Embrace the future and recognize the growing demand for a wide range of fuels or ignore reality and slowly – but surely – be left behind.**

Mike Bowlin, chairman and CEO of ARCO (now BP), 1999

exploration vessels now have the ability to hold position in ocean currents and rough seas with centimetre accuracy and to drill boreholes that are not only thousands of metres deep but are also accurately steered horizontally to reach every pocket of trapped oil and gas.

The rewards are high, but so are the risks, and as offshore oil exploration moves into ever deeper waters, those risks increase. The greatest danger is a blowout: the sudden release of a high-pressure pocket of gas or oil. In theory, oil wells are equipped with elaborate and expensive blowout preventers on the seafloor above the borehole, but, in the Gulf of Mexico in April 2010, a blowout preventer failed. The Deepwater Horizon exploration vessel took the full force of the explosion and sank, killing 11 people and leaving the uncapped well spewing millions of tons of crude oil into the delicate marine environment.

Peak oil 'Peak oil' is the point in time when the maximum rate of global oil extraction is reached, after which the rate of production enters terminal decline. We have probably already reached the maximum rate at which conventional oil and gas can be extracted, but, as fuel prices rise, so more marginal deposits become economical, among them tar sands and oil shale where the hydrocarbons are stuck in the rock and cannot simply be pumped out. In such cases, if they can't be dug out and processed on the surface, the rocks are 'fracked' – fractured by high-pressure fluids so that they become permeable and steam or other solvents can then be used to force them out. But such processes are environmentally unpopular and they use a great deal of energy in the recovery process.

gas hydrates

One significant source of hydrocarbons has not yet been widely exploited: gas hydrates. They are a form of water ice in which large quantities of natural gas, usually methane, are locked within the cages of the crystal lattice. They are only stable under high pressure and at low temperature, and that means that they are only found on or under the seafloor. They may represent twice as much carbon as in all other fossil fuel stores put together. Methane hydrate ice can contain 164 times its own volume of gas, making it tricky to recover without explosive expansion. In the past, warm ocean currents and lowering sea level may have destabilized gas hydrates, leading to big releases and resulting in climate change.

When fossil fuels start to run dry, we will be left with a combination of two alternatives: energy derived ultimately from the Sun (either directly or through biomass, wind and waves) and nuclear power. Historically, nuclear power stations have been fuelled with uranium, which is itself a finite geological resource, though one that could last for several centuries. The reason for this is probably that the early nuclear programme also provided plutonium for military purposes. A potential and more abundant alternative could be thorium, providing a resource that would last a millennium. Ultimately, nuclear fusion, the process that powers the sun, may be the only option.

the condensed idea
Past life, present fuel

21 Riches from the deep

Everything we manufacture, from motor cars to mobile phones, contains materials mined from the Earth. But how did they get there? How can we find them and extract them? And are there enough to go round our increasingly technological civilization?

Although the bulk composition of the Earth's crust may be silicate rock, that approximation overlooks a wealth of minerals containing every element in the periodic table down to uranium. If the rocks were evenly mixed, the concentrations would be so low that it would be very difficult to purify and make use of any of them. Fortunately a number of processes have refined useful minerals into economically extractable concentrations in rocks known as ores. The strict definition of an ore is a mineralized rock that can be profitably mined; but such are the vagaries of market prices, environmental factors and politics that the word is commonly used to refer to any concentration of a potentially useful mineral.

Crustal cooker The majority of mineral deposits are associated with the emplacement of hot rock, or magma. Sometimes, mineral deposits can form within the magma itself. As a molten magma such as granite cools, different minerals crystallize out at different times and may settle out in layers. Also, some of the still molten liquids may not be able to mix with each other. That is the case with some sulphur-rich liquids containing nickel, copper and platinum, which may separate out at the base of a magma chamber.

timeline a brief history of mining

43 ka	5,500 BC	3000 BC	2500 BC
First evidence of mining of haematite for ochre in Swaziland, South Africa	Copper mining begins in the Balkans	Neolithic flint mine starts in Grimes Graves, Norfolk	Bronze (an alloy of copper and tin) comes into use, possibly including tin mined in Cornwall

Most of the world's ore deposits are associated with a magma body. Deposits may originate from magma but build up in the rocks around it. The key to many of these is water. As hot magma rises and the pressure on it is released, water previously dissolved in the magma is expelled into cracks in the surrounding country rock.

Hot water Hydrothermal fluids carry with them many salts, in particular sulphides, which can form complexes with many metals in solution. These minerals come out of solution as the temperature and pressure lower, coating cracks in the rock and turning them into mineral veins. These can be on any scale from microscopic up to several metres thick. Other, less valuable minerals that often accompany ores in veins are known as gangue minerals and can guide miners to the richest veins.

Rare earth

Our increasingly technological civilization is coming to rely more and more on certain elements, some of which are inherently rare, hard to extract or in short supply. For example, high-performance magnets, lasers, solar cells, special glass, displays and touchscreens all use a variety of unusual elements. Globally, the rarest are the platinum group of metals, which are in very short supply. The so-called rare earth elements, such as neodymium, used for the magnets in wind turbines, are not so rare but are only mined in a few places. A few countries, notably China, control the bulk of the world's supply. In the future, new sources need to be found: both for conventional mining and to develop new ways to extract these elements – perhaps from seawater and from recycled waste. Ultimately, we may need to look to space and mine asteroids, which are particularly rich in some of these metals.

1400 BC	700 BC	AD 100	1709	2007
Iron dagger deposited as treasured possession in Tutankhamun's tomb	Iron Age comes to Britain	Widespread mining throughout the Roman Empire	Abraham Darby builds the first blast furnace fuelled by coke	Chuquicamata copper mine in Chile becomes the most productive in the world

The hydrothermal brine can be highly acidic and can react with the surrounding country rock, sometimes dissolving it and replacing it with minerals. These are sometimes rich in gold, silver and copper. Where granite magma comes into contact with carbonate rocks such as limestone, there can be extensive chemical reactions producing what is known as a skarn deposit, which may be rich in iron, copper, lead, zinc and tin.

Nearer the surface, groundwater in the surrounding rock can itself become heated and infused with minerals. This is likely to be lower in sulphide minerals and less acidic, but can still deposit gold, silver, copper, lead and zinc. These are known as epithermal deposits.

Treasure from the mid-ocean Seawater percolates through much of the rock that makes up the ocean crust. Where it meets hot or cooling magma, it starts to dissolve minerals from the rock as it is conveyed through cracks and fissures to emerge on the ocean floor as hydrothermal vents or black smokers. By now, the water is rich in sulphide minerals including lead, zinc and copper. As it vents into the ocean, the hot water is suddenly cooled and can no longer dissolve all the minerals which precipitate out as the 'black smoke'. This builds up around the vent as a series of delicate chimneys that collapse to form thick deposits of sulphide minerals. Often these will eventually be subducted back down into the

> **Where there is cinnabar above, yellow gold will be found below. Where there is lodestone above, copper and gold be found below. Where there is calamine above, lead, tin, and red copper will be found below. Where there is haematite above, iron will be found below. Thus it can be seen that mountains are full of riches.**

Guan Zhong, *c.*720–645 BC

gold rush

Nature can sometimes assist with the mining process. Gold is one of the few metals found in its native, chemically pure form. But it is often sparsely distributed through huge quantities of rock. Fortunately, as those rocks erode, rivers and streams sift the debris and can concentrate the dense gold particles into so-called placer deposits. Prospectors are essentially continuing that process when they pan for gold in sands and gravels. It was placer deposits that were the target of the famous Californian and Klondike gold rushes of the 19th century.

Earth with the old ocean crust. But occasionally they are preserved on land, as is the case in Cyprus, where the copper sulphide deposits have been mined since the Bronze Age.

Rich soil The final type of ore deposit can arise far from magmatic heat. All it takes is a thick layer of soil in a warm, wet, tropical climate. These conditions can produce acidic waters in the topsoil which cause chemical erosion, dissolving and removing many soil minerals and concentrating those rich in certain metals. That is how the iron deposits in the Weald of Kent were produced 100 million years ago, and it is how deposits of laterite are still being formed in the tropics today. Bauxite is a form of laterite rich in aluminium and is still the main source of that element.

the condensed idea
Intrusive heat and water bring mineral wealth

22 Deep diamond secrets

Diamonds are not only a girl's best friend. Geologists are pretty keen on them, too. Locked in the crystal lattice of this, the hardest mineral, can be secrets from their 3-billion-year history and their journey through the mantle of the Earth.

Diamonds are made of carbon, just like soot or graphite. But in this case, high pressure has reformed the chemical bonds into a three-dimensional lattice that makes them incredibly hard, transparent and, when cut in the right way, alluringly sparkly. Occasionally, a few nitrogen atoms substitute for carbon in the lattice, giving a yellowish diamond. Boron leads to a blue diamond. Damage due to radiation can make a diamond green, while shear stress can produce brown, pink and even red diamonds.

Growing diamonds For carbon to form into diamond it needs very high pressure – typically the pressures encountered between 130 and 200 kilometres (81 and 124 miles) underground. But the most favourable temperatures for the process, 900 to 1,300 degrees Celsius, are relatively cool for such depths. The right combination of pressure and temperature exists in the lithosphere at the roots of ancient continents, where most diamonds mined on Earth are to be found.

Supersonic eruptions To get diamonds to the surface from such depths requires unusual events to take place. Most volcanoes have their roots, where melting occurs, perhaps between 5 and 50 kilometres (3 and

timeline the long, deep history of a Brazilian diamond

2.5 Ga	2.2 Ga	1.3 Ga
Micro-organisms absorb carbon dioxide, die and accumulate in sediment	Carbon-rich sediment is subducted down into the mantle with old ocean lithosphere	Deep in the lower mantle, carbon begins to crystallize out as diamond

Emerald

Though not as hard or valuable as diamond, emeralds are still both hard and valuable. They are a form of the silicate mineral beryl, coloured by traces of chromium or vanadium. Beryl is frequently found in pegmatites – the final fraction of a granite intrusion to crystallize. The crystals can be very large: a beryl crystal 18 metres (59 ft) long was found in Madagascar. Where hydrothermal fluids contain chromium and vanadium, beryl can grow as green emerald. That is how most deposits were formed. But in Colombia, particularly fine emeralds are found in black shale, where tectonic pressure has squeezed chromium-rich water through the rocks. Each source of emerald has its own oxygen isotope signature. This helped archaeologists discover that a Roman emerald earring originated in the Swat Valley in Pakistan!

31 miles) beneath the surface. But kimberlite volcanoes go much deeper, sourcing their magma from the depths where diamonds lie. No one in recorded history has witnessed a kimberlite volcano erupting. It would be spectacular, but you wouldn't want to get too close. Because the magma comes from such depths, releasing it at the surface is like taking the cork out of a shaken-up bottle of champagne. Hot magma can be ejected at supersonic velocities.

Diamonds are not usually found coming out of the top of such volcanoes, or if they once did so, they have long since vanished. The biggest diamond mines in the world tend to be in the volcanic pipe that once fed the volcano. These can be several kilometres deep and hundreds of metres wide, shaped a bit like a carrot. In the case of the famous mines of Kimberley in South Africa (which gives its name to the kimberlite

200 Ma	**100 Ma**	**Now**
A mantle plume carries the diamond to the upper mantle	Diamonds are erupted on the surface in a kimberlite volcano	Diamonds are mined, cut and polished and used to adorn our bodies

volcano), the eruptions took place 100 million years ago and erosion has taken the ground level down to about a kilometre or more deep in the pipe. The diamonds themselves are mostly over a billion years old and sometimes over 3 billion. Not all kimberlite volcanoes contain diamonds and not all of those that do are worth mining. It can be necessary to crush many tons of hard rock in order to find each diamond, but their value makes it worthwhile.

Organic diamond The carbon that forms into diamond can come from one of two main sources, distinguished by the ratio of different isotopes of carbon: carbon-12 and carbon-13. Carbon that has originated within the Earth's mantle tends to contain more carbon-13. But some diamonds contain isotopically light carbon, more like that found in living organisms in the sea. The conclusion is that this carbon has indeed been through the carbon cycle of living organisms, becoming incorporated in sediment on the ocean floor and then subducted back down into the mantle to form diamond.

Message in a diamond Jewellers like a perfect, flawless diamond, but geologists love the blemishes. They can represent inclusions of material that was present when and where the diamond formed and can

cutting diamond

As it is the hardest substance known to man, diamond is very difficult to cut! Fortunately, it is not equally hard in all directions, so, once the crystal axes are determined, it is possible to cut and polish facets on a diamond using saws or grindstones edged with other small diamonds. People have been cutting diamond since the 14th century, but today it is a high-tech industry. Ninety per cent of the world's diamonds are cut in Surat, in the Indian state of Gujarat. Often, a computer model of the rough diamond is created to determine the crystal axes and inclusions and work out how to get the most valuable cut diamonds from it. Increasingly, lasers are also used to help with the cutting.

‘A diamond is a chunk of coal that is made good under pressure.’

Henry Kissinger

often be used both to date that stage in its formation and to reveal how deep in the Earth it was at that time. Thus it is possible to recreate the life story of a diamond.

Prof. Steve Shirey and colleagues at the Carnegie Institution in Washington have sliced into thousands of diamonds over the years to sample their inclusions. Recently, they noticed that all the diamonds older than 3.5 billion years only contained mantle carbon. Carbon from organic sources in ocean sediments only turns up later. They concluded that this must mark the start of the first ocean crust subducted and the onset of the Wilson cycle that drives continental drift.

Deep diamonds Occasionally, diamonds turn up that carry in their inclusions the mineral signatures of much greater depths than the continental lithosphere. Some even contain the high-pressure perovskite minerals from the lower mantle. Diamonds from one mine in Brazil not only carry the signatures of the lower mantle, deeper than 660 kilometres (410 miles), but also the light carbon isotopes of organic carbon. This provides one of the first pieces of direct evidence of whole mantle circulation: ocean crust with carbon-rich sediments that has been subducted down to the base of the lower mantle and eventually contributed material to a mantle plume that erupted in what is now Brazil during the Cretaceous period (see timeline).

the condensed idea
History of a crystal under pressure

23 The rock cycle

No rock and no continent is an island entire of itself. In this section we come above ground to include the processes at work in air and water: atmosphere and ocean. The idea that sums it all up is that of the rock cycle. What comes up eventually goes down again. Planet Earth is the ultimate recycling centre.

We have already heard how the solid mantle circulates and a fraction of it melts and creates the crust. We have heard how ocean crust subducts back down into the mantle, taking some of its accumulated sediments with it. Now we come to what happens in between – on land and under the sea. Air, water, heat, even life itself goes in cycles, and they are circulations we will come back to. Here we examine the cycles affecting the substance of the Earth itself: the rocks.

Hutton's big idea One of the first to recognize the cyclical processes at work on the continents and in the oceans was James Hutton, sometimes referred to as the father of modern geology. In 1785, he first described sediment erosion on land, transport to the ocean, accumulation on the ocean floor, hardening into rock and subsequent uplift for erosion to begin again. He was ahead of his time in that he realized the cycle must include processes not easy to observe at the surface, and that it required long periods of time – longer than those suggested by theologians.

Plutonism Hutton's idea of the rock cycle was closely linked with his belief in plutonism: the idea that many rocks, such as basalt and granite, had once been molten magma. This was disputed by the rival theory of neptunism, which held that all rocks were sedimentary and had been laid down by water. Hutton was the first to suggest that, if they became buried

timeline

1679	1776	1779	1788
Robert Hooke shows that fossiliferous layers are too thick to have formed in the 150 days of Noah's flood.	James Keir claims the Giant's Causeway formed when molten rock cooled	Comte de Buffon suggests that the Earth is at least 75,000 years old	James Hutton publishes *Theory of the Earth*

JAMES HUTTON 1726–97

After qualifying in medicine, James Hutton returned to his native Edinburgh and took up farming nearby. At the age of 42, he sold his now prosperous farm and returned to the city, where he was an active member of the Edinburgh Philosophical Society, which later became the Royal Society of Edinburgh. Here he studied geology and chemistry and in 1788 published his classic work *Theory of the Earth*.

In it, he established the concept of the geological cycle and the immense amount of time required for gradual processes to complete it. He also recognized the role of heat and pressure and became a convinced plutonist, suggesting that rocks such as granite were produced from molten magma and not through sedimentation.

deep enough, rocks of all three types – sedimentary, igneous and metamorphic – would melt. He further proposed that the molten rock would rise to erupt from volcanoes or intrude into shallower rocks and produce mountain ranges.

Unlike on the dry, airless surface of the Moon, rocks on the surface of the Earth are seldom in equilibrium with their environment. No sooner have they been lifted into mountain ranges but the forces of air, ice and water begin to erode them back down again. We will look at the results of those physical mechanisms in more detail in the next section.

Chemistry in the rock cycle One of the key processes in the rock cycle is not physical but chemical. Carbon dioxide dissolved in rainwater makes a weak acid that reacts with the minerals particularly in rocks such as basalt to produce the minerals of clay. They incorporate water into their structure which goes on to lubricate some of the tectonic processes later in the cycle. Such chemical erosion also changes the composition of the atmosphere, so that the uplift of new mountain ranges is closely followed by a drawdown of carbon dioxide from the atmosphere.

1797	1807	1830	1964
Sir James Hall proves that igneous rock can crystallize from molten material	The Geological Society of London becomes the first devoted to the new subject	Charles Lyell's *Principles of Geology* shows Earth must be hundreds of millions of years old	Tuzo Wilson extends the rock cycle to include plate tectonics

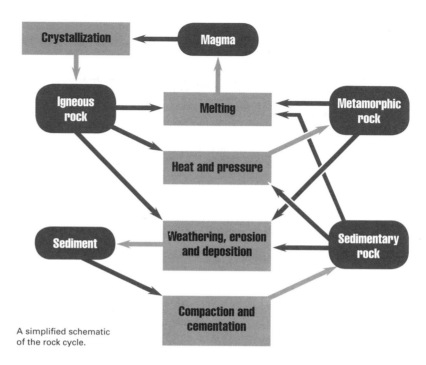

A simplified schematic of the rock cycle.

Water is essential to the rock cycle. It dissolves carbon dioxide to produce the carbonic acid for chemical weathering. It erodes soft sediment and dissolves soluble minerals. It forms ice that expands in the cracks and shatters the rocks, and it builds into glaciers that grind the rocks down. Water transports sediments to places where they can accumulate and subsequently lubricates their passage into the Earth and lowers the melting point of the magmas they produce.

Turning to stone The definition of a 'rock' is as loose as some of the sediments that are referred to as rock! In practice, the transition from the mud, sand and gravel that washes down from erosion and ultimately forms into a sedimentary rock can be slow and gradual. Essentially, the process represents the collapse of the

Unconformity

Hutton's ideas were founded on observation, and he visited several sites in Scotland now known as Hutton's Unconformities. At the first, on the Isle of Arran, Cambrian schists are deformed so that the strata are almost vertical. They have then been eroded and covered by horizontal layers of much younger sandstone. Hutton later found a much clearer example near Jedburgh and wrote how he had 'rejoiced at my good fortune in stumbling upon an object so interesting in the natural history of the Earth'. Such unconformities proved that there had been successive cycles of uplift, erosion and subsequent deposition.

pore structure within the rock, either through compaction by the weight of sediment above, or through cementation as the pores are filled up with new chemicals that bind the grains together. The chemical cement can come from within the sediment itself, or it can percolate in from another source, in which case the process is known as diagenesis.

Of course, the rock cycle isn't a simple circle. All three basic types of rock – sedimentary, igneous and metamorphic – can get uplifted and eroded. All three can become buried and changed by heat and pressure. And all three can undergo partial melting.

Wilson cycle In the 1960s, Tuzo Wilson developed the rock cycle further to incorporate his new ideas of plate tectonics. He incorporated the idea of mantle convection into the sequence, deepening and completing the cycle through subduction and magma generation in the mantle.

the condensed idea
Rocks in an unending cycle

24 Carving landscape

Every stone has an internal story to tell of its formation, composition and history. Rocks and the landscape in which they are found also tell an external tale of the forces that erode and sculpt them into the landforms we see around us. Physical geography tells the story of landscape.

Solar power Ultimately, almost all the processes of erosion on the surface of the Earth are powered by the Sun. It is solar energy that circulates the atmosphere and drives the wind. Sunshine evaporates water, which forms clouds and falls as rain or snow, feeding rivers and glaciers. What the Sun lifts up, gravity pulls down, adding cutting power to falling water, and toppling rocks and taking the debris to the lowest points in valleys, basins and oceans.

The rate of erosion The higher they rise, the quicker they fall. Erosion rates are usually much faster in young mountainous regions. The rate depends fundamentally on two factors: the rate of weathering and the rate of transport. Weathering can itself be of two types: physical and chemical. In the process of chemical weathering, weakly acidic rainwater dissolves rocks such as limestone or reacts with silicates to produce clays. This tends to happen along cracks and fissures and can, in turn, loosen larger fragments for physical erosion.

Water power The most powerful agent of physical weathering is water, particularly when it turns to ice. As water freezes, it expands, so ice in narrow cracks can act like a wedge and shatter hard rock. On a larger

timeline the Grand Canyon

2 Ga	1 Ga	500 Ma	280 Ma	230 Ma
Oldest rocks in the canyon	Start of 'the great unconformity' with the region on land, suffering erosion	End of the unconformity. Marine sediments return	Permian period. Windblown sand dunes deposited on land	Youngest sedimentary limestone in the canyon

scale, ice in a glacier has tremendous power, gouging out broad valleys and grinding rock into mud. Fast-flowing mountain streams can transport debris as quickly as it forms, so erosion is limited by the rate of weathering, producing a barren landscape with a lot of exposed rock.

Where erosion is limited by the rate of transport, sediments tend to accumulate. Such was the rate of weathering during the last ice age that transport couldn't keep up. A large proportion of the sediment load in many rivers in northern latitudes today is made up of loosely consolidated material left by the retreating ice and still soft enough to be quickly eroded.

> **'Physical geography and geology are inseparable scientific twins.'**
>
> **Sir Roderick Impey Murchison, anniversary meeting of the Royal Geographical Society, 1857**

The story of a grain of sand

Lifetimes have been devoted to the study of grains of sand! Any simple statement suggesting, for example, that rounded grains are windblown and angular grains are carried by water is probably wrong. The roundedness of sand grains is largely a matter of how long they've been bouncing around. However, windblown grains tend to get rounded more quickly and, under the microscope, have a finely pitted or frosted surface. Water tends to cushion the collisions. Wind is also better at sorting the grains, blowing fine grains further than coarse ones. By measuring the length of time quartz grains have been exposed to cosmic rays on the Earth's surface, it has been possible to estimate that sand grains take a million years to cross the Namib Desert in southern Africa.

17 Ma	5.3 Ma	3.2 ka	1919
Possible date for the start of formation of the West Canyon	Gulf of California opened, rapid deepening of the canyon. East and West canyons connect	Pueblo people occupy the region	Grand Canyon declared a national park

Soil Wherever sediment accumulates on land, it can lead to the development of soil that may support vegetation. That will, in turn, reduce erosion by stabilizing the soil and reducing the power of raindrops and running water. But, at the same time, roots can break up subsoil or loosely consolidated rock, while decaying organic matter can produce humic acids, which enhance chemical weathering.

Once vegetation is cleared, by fire or farming, erosion can increase dramatically. This was the case in parts of the United States in the 1920s and 1930s, when attempts at intensive agriculture in marginal lands led to dustbowl conditions and the loss of billions of tons of soil.

Wind power In arid lands where there is neither water nor vegetation to protect the rocks, wind becomes the dominant player, picking up grains of sand and using them like tiny chisels to carve a characteristic desert landscape. The wind may not have the power to keep individual sand grains airborne for long. Instead, they move in a process called saltation, bouncing along in short hops. But each time one hits the ground, it knocks others into the air. In this way, wind erosion can be greatest near the ground, leading to the undercutting of rock formations.

The Grand Canyon

The Grand Canyon in the USA is probably the most spectacular erosional feature on Earth. In spite of claims by religious fundamentalists that it was caused by the biblical flood, geologists agree that it is millions of years old. Dates from calcite cave deposits suggest it may have already started forming 17 million years ago. It is 446 kilometres (277 miles) long, nearly 30 kilometres (19 miles) wide and 1,800 metres (5,906 ft) deep, slicing through nearly 2 billion years of sediments. Carved by the Colorado River, it is a spectacular demonstration of the cutting power of water.

Landscape features Erosion may begin on the scale of raindrops and grains of sand, but the characteristic features it carves in the landscape are altogether grander. There are classic differences between mountainous landscapes carved by ice or water. Glaciers carve out broad, deep U-shaped valleys with concave sides and only gentle bends. Where smaller glaciers flow into them, it is the surface of the ice and not the base of the valley that lines up, leaving hanging valleys that begin high up the sides of larger valleys. A glacial valley may be over-deepened – ice can flow uphill – and that can leave lakes and fjords in the valley floor.

> **The elements that unite to make the Grand Canyon the most sublime spectacle in nature are multifarious and exceedingly diverse.**
>
> **John Wesley Powell, after making the first recorded trip along the Grand Canyon in 1869**

River valleys, by contrast, are V-shaped, with straight or convex sides. Water can turn corners, so there may be tight bends in a river valley with overlapping spurs from each side. There may be lakes and waterfalls caused by geological features, but water always runs downhill so it will not over-deepen its course.

The evolution of landscape In 1899, the American geologist William Morris Davis published his cycle of erosion, in which he proposed that uplift of a landscape was followed by stages he called youth, maturity and old age. His youthful landscape had high peaks and narrow, deep valleys. By maturity, the valleys have widened; by old age, they have become low-lying plains. Though it still appears in textbooks, that cycle is now thought to be an oversimplification. Each feature of the landscape is a product of the underlying geology and the forces at work on it, and each has a unique story to tell.

the condensed idea
What goes up must come down

25 Gradualism and catastrophism

The Earth beneath our feet seems solid and unchanging, but, taken over geological time, gradual changes might move mountains. More dramatic are the changes wrought by hurricanes and earthquakes, floods and volcanoes. So is the planet moulded by catastrophe or gradual change? It's a debate that raged in the 18th century and has not yet concluded.

Biblical catastrophe Prior to the 18th century, any scientist would have trained initially in theology. The church held the academic high ground, and it supported Bishop Ussher's calculation, from the generations described in the Bible, that the Earth had been created in 4004 BC. Six thousand years was simply not enough time for the everyday processes of weathering and transport to have deposited all the rocks and carved the landscape. The only alternative was that something far more catastrophic had happened, and that idea gained currency from biblical stories such as that of Noah's flood.

The foremost advocate of catastrophism was the French baron Georges Cuvier (1769–1832). He was an anatomist and had not studied field geology beyond the Paris basin, but he was nevertheless impressed by how different layers of rock contain different fossils and how some layers are tilted with respect to others. Over biblical timescales, he argued, it was impossible for such changes to have happened without violent upheaval. But he proposed no mechanism nor any reason why such catastrophes might have taken place.

timeline

1654	1787	1796
Bishop Ussher proposes that the Earth was created in 4004 BC	Abraham Werner proposes his neptunist theory	Baron Cuvier proposes his principles of catastrophism

Neptunism Baron Cuvier was also a supporter of the neptunist idea, proposed in Germany by Abraham Werner and supported by the writer Goethe, that all rocks, including basalt and granite, were precipitated from the water in some primordial ocean. They recognized that fossils, sometimes found high on mountains, were often of marine organisms and argued that there must once have been a vast, deep ocean. Perhaps the whole of Earth was made of water, from which solids precipitated.

> **'The present is the key to the past.'**
> **Sir Archibald Geikie, 1865**

Uniformity The principle of uniformity, as laid down in Hutton's great work *Theory of the Earth* (1788), is the nearest thing the geologists have to a fundamental law. Its essence was summed up succinctly some years later by another Scottish geologist, Sir Archibald Geikie, in the phrase 'the present is the key to the past'. In other words, all the changes and processes in the geological record can be brought about by processes at work on the Earth today.

BARON GEORGES CUVIER 1769–1832

Following his arrival in Paris in 1795, Cuvier published a paper comparing the skulls of African and Indian elephants with fossil remains of a mammoth and something then only known as the Ohio animal – now identified as a mastodon. He recognized that they were distinct species and that the fossil ones were now extinct. The possibility of extinction of species had not previously been widely accepted. Because his anatomical studies showed clear differences between species, with no intermediate forms, Cuvier rejected the idea of evolution. He thought that extinction and formal change could only be brought about by catastrophe and championed that theory against uniformitarian arguments.

1788
John Hutton publishes
Theory of the Earth,
proposing gradualism
and plutonism

1830
Charles Lyell publishes his
Principles of Geology

1865
Sir Archibald Geikie coins
the phrase 'the present is
the key to the past'

'We find no vestige of a beginning, and no prospect of an end.'

James Hutton, *Theory of the Earth,* **1788**

At first glance, nothing seems more constant and durable than rock. But look more closely: that patch of mud washed out by last night's rain, that line of sand left behind at high tide – gradual change is all around us. What it needs to transform the Earth is time: millions and millions of years of time. Once you escape the confines of Bishop Ussher's time limits, everything becomes possible.

Gradualism Uniformitarianism or gradualism was the theme taken up by another great British geologist, Sir Charles Lyell. The subtitle of his great 1830 work *Principles of Geology* made it very clear which argument he supported: *being an attempt to explain the former changes of the Earth's surface, by reference to causes now in operation.* Lyell's insistence on the length of time needed to support gradual change also provided the backdrop for his friend Charles Darwin to develop his theory of evolution by natural selection.

Neo-catastrophism Gradual change over geological time can explain many things. But every day is not the same. One day is fine; the next it rains. Every few years there might be a catastrophic storm or a terrible flood. Earth tremors and volcanic activity go on all the time, but every few months we hear of a truly devastating earthquake and every few thousand years a supervolcano erupts. Small events are common and big events are rare, but they still happen. Sometimes they are exaggerated in Hollywood films or even science documentaries, but catastrophes still have their part to play. They leave more trace in their brief hour of destruction than the ages that pass unrecorded between them.

If we look back through geological time, things have certainly not been uniform and change has been very far from gradual. In the past, the climate and even the composition of the atmosphere have been different. The Earth has been cooling down and the Sun warming up. Life has transformed the seas, moved on to the land and started to concrete it over! Gradual change certainly continues, but the initial conditions are not as they were a billion years ago.

Extinction It is also clear from the fossil record that there have been some sudden, major catastrophes resulting in the loss of a third or even half of all species on Earth. Whether the cause was asteroid impacts, volcanic eruptions, magnetic reversals, bombardment with cosmic rays or climate change, these extinction events would clearly signify catastrophes for the creatures around at the time. As the geologist Derek Ager put it, it's like the life of a soldier: long periods of boredom and short periods of terror.

the condensed idea
The present is the key to the past

26 Sedimentation

They are only a thin veneer over the Earth's surface, making up less than 10 per cent of crustal rock, but sedimentary rocks are the ones we encounter most frequently and most often use to build our homes and our roads. Each carries the story of its formation, which can be read like a book of stone. After igneous and metamorphic rocks, this is the third main rock type.

The formation of sedimentary rock is the endpoint in the rock cycle – if a cycle can be said to end. The ground-up or dissolved remains of mountain chains and even the bodies of deceased plants and animals all go to make up sedimentary rock.

Types of sediment Sedimentary rocks are normally classified by their source, composition and texture. By far the most abundant are clastic rocks, that is, rocks made up of grains or fragments from their eroded source. These are mostly silicates and often predominantly quartz, the most resistant common source material. Feldspar is also a common component, as are clay minerals derived from chemical weathering of basalt.

Non-clastic sedimentary rocks can be organic or chemical. Organic examples include lignite, coal and oil, shell rock and also some limestones made of the skeletons of calcareous microorganisms. Chemical examples include rock salt, gypsum, anhydrite and limestone deposits precipitated in shallow seas or in caves.

timeline the sediments beneath London

420 Ma	130 Ma	70 Ma
Ancient Silurian rocks deep beneath the capital	Gault clay deposited in deeper water lacking oxygen	Very thick deposits of white chalk laid down in a warm shallow sea

classifying clastics

Clastic sedimentary rocks are classified by the size of the grains within them. Gravel is classified as grains above 2 millimetres in size; sand from 2 to 0.065 millimetres; and mud those grains smaller than that. Mud can be further subdivided into silt and, for the finest grains of all, clay. Gravel cemented into rock is known as conglomerate if the pebbles are rounded, breccia if they are angular. Consolidated sand is sandstone; consolidated mud, mudstone. Sand and gravel can be further divided into coarse, medium and fine; and by composition depending on the amount of silica, feldspar or organic matter.

Sedimentary environments Sedimentary rocks can also tell us about the environment in which they formed. One of the parameters is the energy of the environment. A fast-flowing river or a beach with crashing surf is a high-energy environment. A stagnant lake, mud flat or ocean deep is a low-energy environment. Particles of all sizes can be transported in a high-energy environment. As the energy falls, for example as a river slows on its flood plain, it can no longer transport the larger grains, and first gravels, then sands and finally mud will fall out of suspension.

Some sedimentary rocks are deposited on land. The commonest are windblown, which are very well sorted, and glacial, which can contain particles of all sizes from large boulders to the finest clay. Other terrestrial deposits include peat, which can be consolidated and compressed into lignite or even coal.

> **❛I am convinced, by repeated observation, that marbles, limestones, chalks, marls, clays, sand, and almost all terrestrial substances, wherever situated, are full of shells and other spoils of the ocean.❜**
>
> **Comte Georges-Louis Leclerc de Buffon, 1749**

55 Ma	45 Ma	0.5 Ma
A sequence of estuarine, lake and marine deposits of sand and silt are laid down	Thick deposits of London clay form in shallow sea	Advancing ice diverts the Thames to its present valley, where it leaves gravel deposits

Deltas and dunes Rivers deposit huge quantities of sediment when they reach the sea and their energy drops. A river delta can spread out over hundreds of square miles and build up sediment thousands of metres thick. Normally, sediments are deposited in horizontal layers, but not always. Where they are eroded and redeposited, the layers can be at a low angle down the face of the slope. This can occur both in river deltas and on land in dunes and can produce successive bands of angled beds known as cross-bedding.

Transgressions The sea level can rise or fall and land can be uplifted or sink. The result can be a changing sequence of sedimentary environments, reflected in different rock layers. Where the coastline is retreating inland and the deposits are from progressively deeper waters, it is known as a marine transgression. Where the sea is retreating and deposits are getting shallower, it is a regression. If a landlocked sea dries out altogether, it leaves a layer of chemical evaporite deposits such as rock salt and gypsum.

Sedimentary basins Sometimes tectonic forces can stretch the Earth's crust, causing it to thin. That, in turn, causes the crust to sag down, allowing the sea to flood over it and deposit sediments. The weight of those may cause the basin to sag further in a vicious cycle in which sediment layers up to 10 kilometres (6 miles) thick can develop.

Diagenesis

Once a sediment is deposited, that is not the end of the story. As layer builds upon layer, the sediments get squeezed and compacted and cemented in a process called diagenesis. Clay contains as much as 60 per cent water by volume and, as that is squeezed out, the clay often compacts into a finely layered shale. If there is calcium carbonate present, it will cement the clay into a hard calcareous mudstone. Squeezing sand will tend to cause the sand grains to dissolve where they touch and redeposit in the spaces, forming a hard sandstone.

> **Rocks are records of events that took place at the time they formed. They are books. They have a different vocabulary, a different alphabet, but you learn how to read them.**
>
> **John McPhee**

Because the cause is stretching, the lithosphere also thins, bringing the hot asthenosphere closer to the sediments and cooking them, a process that helps in the maturation of oil and other hydrocarbons. A good example of this is the North Sea.

Another type of sedimentary basin is associated with ocean-floor subduction. As ocean crust dives down beneath the margin of a continent, sediment may get scraped off to form an accretionary prism, trapping a shallow basin behind it in which sediments from the continent accumulate. That is known as a fore-arc basin. At the same time, volcanic peaks in the continental margin depress the continent, resulting in a shallow basin with accumulating sediment behind the mountain range – a back-arc basin.

Deep-sea sediments In the deep ocean, far from land, sedimentation may be very much slower. Below 4,000 metres (13,123 ft), calcium carbonate is dissolved by the pressurized water so that limestone cannot form in the deep ocean and the skeletons of calcareous microorganisms never make it to the deep ocean floor. But skeletons based on silica can and do.

the condensed idea
Laying down history one layer at a time

27 Ocean circulation

Covering 71 per cent of the planet's surface and containing 97 per cent of all its water, oceans are central to understanding the dynamic processes on the surface of the Earth. The top 2 metres (6½ feet) of ocean water contain more heat than the entire atmosphere, and the circulation of that heat in ocean currents performs a vital role in controlling and moderating global climate.

Coriolis force Surface ocean currents are driven in part by the prevailing wind. But, according to Newton's laws of motion, any moving mass will try to continue in a straight line. On the surface of a rotating globe it is unable to do so, but it will still try. The result is the Coriolis force. In the northern hemisphere, the force makes a moving current tend to pull to the right, while in the southern hemisphere, it will pull to the left. The result is an ocean gyre: a vast system of ocean currents going round in circles. The best example is in the northern Pacific, where there is no land to impede it. It is a sad reflection on our disposable plastic age that a vast raft of floating plastic waste has accumulated in the calm centre of the gyre.

> **How inappropriate to call this planet Earth when it is quite clearly Ocean.**
>
> **Arthur C. Clarke**

In January 1992, during a Pacific storm, three container-loads of plastic ducks from a Chinese factory were washed overboard. About two-thirds of the 29,000 ducks drifted south and landed on Indonesian

timeline milestones in oceanography

1777	1812	1835	1872–76
British geographer James Rennell suggests that ocean currents are driven by the wind	Alexander von Humboldt describes cold, deep currents flowing from the poles towards the equator	Gustave-Gaspard Coriolis identifies the rotating system of surface ocean currents	HMS *Challenger* makes first voyage dedicated to scientific oceanography

The Younger Dryas

Around 12,800 years ago, the world was slowly emerging from an ice age. Temperate forests were beginning to return to north-west Europe. Then, suddenly, pollen from the Arctic tundra plant dryas appears in sediment cores, marking an abrupt return to cold conditions across north-west Europe and Greenland. The explanation is probably that the Atlantic conveyor shut down due to a sudden influx of freshwater, possibly from a lake of meltwater trapped behind the retreating ice sheet of North America. As a result, the Gulf Stream no longer brought warm water northwards. About 1,200 years later, the cold period ended as quickly as it had begun.

islands and in Australia. The remainder travelled north and entered the so-called North Pacific Gyre. A few made it through the Bering Straits and became trapped in slow-moving Arctic ice, eventually to turn up in the North Atlantic eight years later.

Remote sensing Oceanographers have more sophisticated ways of tracking ocean currents, using floating buoys at the surface and ones which sink to a predefined depth before returning to the surface to radio data back by satellite. The space age has revolutionized oceanography. Once research ships and Merchant Navy volunteers could only monitor temperature and salinity by dropping a bucket over the side of a ship to collect samples; today satellites can monitor waves and currents and even the amount of phytoplankton in the water on a daily basis from space.

Salt For 4 billion years, rain has been falling on the land, washing through the rocks and flowing down to the sea, carrying with it dissolved salt. Over geological time, the sea has been getting more and more salty.

1894	1930s	1940s	1943
Fridtjof Nansen attempts to reach the North Pole in a study of how sea ice drifts	Sir George Deacon develops new ways of measuring deep ocean currents	Henry Stommel studies the Gulf Stream and the formation of Antarctic bottom water	Jacques Yves Cousteau invents the aqualung

Today, if all the oceans evaporated and the salt left behind were spread out evenly, it would form a solid layer 75 metres (246 ft) thick.

Dissolved salt is not divided evenly between the oceans. The Baltic, with lots of freshwater rivers and low evaporation, contains about 5,000 parts per million (ppm) of salt. The Persian Gulf, with high evaporation and few rivers, has about 40,000 ppm. Both salt and temperature change the density of seawater and, together with the effects of wind and the Earth's rotation, contribute to ocean circulation, giving it a third dimension as cold, salty water sinks.

Atlantic conveyor belt The phenomenon is best seen in the North Atlantic, where the Gulf Stream brings warm water north-eastwards from the Gulf of Mexico, keeping the climate of the British Isles temperate when compared to eastern Canada at the same latitude on the other side of the Atlantic. As the Gulf Stream heads north, it cools, and evaporation makes it saltier. So, as it approaches the Arctic, this dense water sinks and returns south as Atlantic bottom water.

There are hints that global warming is slowing the production of Atlantic bottom water, by adding cold freshwater from melting ice, thus diluting the salt, and possibly by warming the surface layer and causing stratification so that it is unable to sink. If this Atlantic conveyor belt

cold currents

warm currents

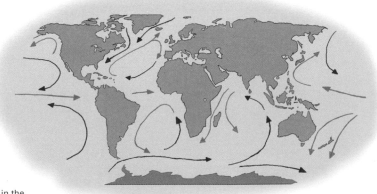

Principal surface currents in the world's oceans.

were to stop, it would be an ironic consequence of global warming that
north-west Europe would suffer much colder winters.

Agulhas leakage Help and salt water may come from the south. The
Agulhas Current in the Indian Ocean gets pinched by the Cape of South
Africa and some of its warm, salty water leaks into the South Atlantic.
Only 10 per cent of the current
leaks, but it still amounts to 200
times the flow of freshwater from
the Amazon River, and it seems to
be increasing with global warming.
It might compensate for Arctic
melting by feeding more salt into
the Atlantic system.

Nutrients The vertical
component of ocean circulation
is also vital for marine life.
Phytoplankton and the marine
creatures that feed off them
love warm, sunlit surface water.
But they quickly use up all the
available nutrients dissolved at the
surface. Deep water offers plenty
of nutrients, but it is too dark for
photosynthesis. Where ocean
currents bring deep water to the
surface, the nutrients come too and
begin to mix with the warm waters,
producing something analogous to the cold fronts in the atmosphere
that feature in weather forecasts. Such zones are a bonanza for plankton.
Sensors on spacecraft tuned to the green colour of chlorophyll reveal
massive seasonal blooms of plankton in these areas.

El Niño

At irregular intervals every five years or so,
atmospheric pressure rises over the Indian Ocean
and Indonesia and drops in the eastern Pacific.
The Pacific Trade winds weaken or blow to the
east, and a current of warm, nutrient-poor tropical
water heads for the coast of Peru. It is called
El Niño (meaning the child), because it arrives
at Christmas, like the Christ child. It blocks the
nutrient-rich Humboldt current, causing South
American fisheries to fail. It brings storms and
rain to desert regions in the western Americas and
drought to Australia and the western Pacific. A year
later, El Niño is often followed by a cold current, La
Niña, producing the opposite effects, with South
American drought and Australian floods.

the condensed idea
Ocean currents: centrally heating the Earth's surface

28 Atmosphere circulation

Spaceship Earth is protected from the ravages of outer space by what, in astronomical terms, is but a fragile veil of air. But from the human perspective, the atmosphere is a vast ocean in which we live, breathe and exist. It is also a great heat engine, driven by solar energy, distributing heat and water vapour around the globe and bringing us weather.

It is hard to say how thick the atmosphere is. It becomes increasingly rarefied as it thins into space. There are still a few atoms around even at the height of the International Space Station, just under 400 kilometres (250 miles) above sea level. Those few atoms are ionized and so energetic that they have a temperature of about 2,000 degrees Celsius, though they are too rarefied to hold much heat. This is the thermosphere.

The protecting veil At a height of about 80 kilometres (50 miles), we come to the mesosphere, which includes the ionosphere, the layer of air ionized by bombardment with cosmic rays from space. This is the fragile veil that protects us from cosmic X-rays. It is also where charged particles from the Sun, streaming in along the magnetic field lines above the poles, produce the spectacular glowing curtains of the aurora. And it reflects shortwave radio signals, enabling international communications.

At 50 kilometres (31 miles), we come to the stratosphere. This is the coldest part of the atmosphere, neither warmed by radiation from above nor convection from below. This is the home of the ozone layer, produced by the action of ultraviolet light from the Sun on oxygen molecules; it

timeline

350 BC	1643	1686	1724
Aristotle coins the term meteorology as the title of a book about Earth sciences	Evangelista Torricelli invents the mercury barometer	Edmund Halley identifies solar heating as the cause of atmospheric circulation	Gabriel Fahrenheit invents a reliable scale for measuring temperature

Monsoon

Land affects atmospheric circulation, both by the thermal effects of its surface and the physical effects of mountain ranges. The greater thermal capacity of the ocean means that, in summer, the air overland heats up more by day and starts to rise, drawing in a sea breeze. The most dramatic manifestation of that is the South Asian monsoon. In winter, the prevailing wind is from the north-east, but between June and September it reverses, pulling in warm, moisture-laden water from the Indian Ocean. The Himalayas force the air to rise and water begins to condense, resulting in up to 10 metres (33 ft) of rainfall in some areas. The Himalayas effectively block the moist air and the Tibetan plateau remains dry. Sediment cores show that the monsoon started at about the time the Himalayas were rising. Less dramatic monsoons also occur in West Africa and East Asia.

screens us from harmful UV radiation. The small amounts of water in the stratosphere can form high ice clouds in polar regions which provide the substrate, catalysed by chlorine compounds released by human activity, for the destruction of ozone.

Where the weather is Eighty per cent of the air and 99 per cent of water vapour are in the troposphere. This varies in thickness between 20 kilometres (12½ miles) over the tropics and 7 kilometres (4½ miles) over the poles. It is separated from the stratosphere by a temperature inversion layer, which prevents much mixing. It is within the troposphere that most of the circulation of heat and water vapour takes place. The troposphere brings us our weather.

Circulating cells Atmospheric circulation in the troposphere can be simplified to a series of circulating cells, transferring heat to high latitudes. The process starts with warm air rising above the equator, or more specifically the zenith line, because it migrates north and south with

1735	**1806**	**1928 and 1933**	**1960**
George Hadley publishes his explanation of atmospheric circulation and the trade winds	Francis Beaufort proposes a scheme for classifying wind speed	Wasaburo Oishi in Japan and Wiley Post in the USA detect the jet stream	First successful weather satellite launched

> **The Sun's rays are the ultimate source of almost every motion which takes place on the surface of the Earth. By their heat are produced all winds … By them the waters of the sea are made to circulate in vapour through the air, and irrigate the land, producing springs and rivers.**
>
> **Sir John Herschel,**
> ***Outlines of Astronomy, 1849***

the seasons, keeping the sun overhead. The air flows north, high in the troposphere, to a latitude of about 30 degrees, by which time it has cooled sufficiently to descend and flows south again, completing the circulation. The same thing happens in the southern hemisphere.

In 1735, George Hadley published a paper showing how what we now know as the Coriolis force makes the moving air always pull slightly to the right, producing the north-east to south-west trade winds in the tropics as it completes its circulation. It took 100 years for his contribution to be recognized; the cell is now known as the Hadley cell. But in fact his explanation was not complete. Were a mass of air simply preserving angular momentum as it moved to high latitudes and hence closer to the Earth's rotation axis, it would accelerate to storm force. In practice, differences in pressure also have an impact, keeping wind speed moderate.

Another system, the polar cells, begins with warm air rising at about 60 degrees north and south of the equator and descending as cold air at the pole. In between the polar cell and the Hadley cell is a third, slightly more complicated circulation known as the Ferrel cell.

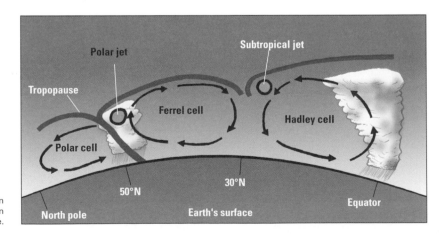

Atmospheric circulation cells in the Northern hemisphere.

Hurricane

Warm, moist air starts to rise above the warm ocean, creating a region of low pressure and pulling more warm, moist air in from the sides. The phenomenon can develop into a vast, spiralling weather system known as a hurricane in the Caribbean, a cyclone in the Indian Ocean and a typhoon in the Western Pacific. They can persist for many days, strengthening and drifting with the trade winds until they strike land. By then, wind speeds can reach 186 mph (300 km/h), accompanied by torrential rain. The low pressure can be so intense that it raises sea level by up to 8 metres (26 ft) in a storm surge. Once over land, hurricanes start to lose their supply of moisture and subside.

These cells not only explain prevailing winds, they also account for important weather systems. Where warm moist air is rising, around the equator and around 60 degrees north and south, it leads to low pressure systems and rain clouds. That is why there are rainforests around the equator and why there is often a run of low pressure systems crossing the North Atlantic. In between, where dry air descends, desert regions are more likely.

Jet stream High in the troposphere where cells meet is a narrow ribbon of air travelling from west to east at high speed. These are the jet streams. There are two in each hemisphere – a polar jet stream and a higher but weaker subtropical jet stream. Wind speeds in the polar jet stream can reach 124 mph (200 km/h). The jet streams can meander in what are known as Rossby waves, which travel more slowly west to east. Their position can determine whether Atlantic rain systems fall on London or Lerwick.

the condensed idea
Wind and weather are driven by heat and moisture

29 Water cycle

One of the features that makes our planet special – perhaps unique – is the presence of water in all three phases: liquid, vapour and ice. This enables it to cycle between ocean, air and earth, transporting heat and bringing life.

Water, or at least its atomic constituents hydrogen and oxygen, are among the most abundant elements in the universe, and yet we know of no other world but our own where there is abundant liquid water on the surface. There may be oceans on Jupiter's moon Europa and Saturn's satellite Enceladus, but they lie beneath many kilometres of ice.

Habitable zone It is possible that a handful of the hundreds of extrasolar planets discovered recently may lie in the habitable zone – the right distance from their parent star to sustain liquid water – but we have yet to prove it. It is of great interest that we do so, because liquid water is one of the few absolute essentials to support life as we know it.

Where the water is It is likely that any water on the surface of the primordial Earth boiled away as a result of volcanism and collisions and was stripped from the early atmosphere by the strength of the solar wind. So all the water on Earth today must have come either from within, via volcanic eruptions, or externally, from comets and asteroids. Today, there are an estimated 1,386 million cubic kilometres (333 million cubic miles) of the stuff, of which about 97 per cent is in the oceans, 2.1 per cent is in ice caps, 0.6 per cent in underground aquifers, 0.02 per cent in lakes and rivers and just 0.001 per cent as clouds and vapour in the atmosphere.

> **When the well's dry, we know the worth of water.**
> **Benjamin Franklin, 1757**

timeline milestones in hydroelectric power

c.250 BC	c.AD 100	c.1300	1878
Philo of Byzantium makes the first recorded reference to a waterwheel	Waterwheels in widespread use in the Roman world for mining and irrigation	Water mills in widespread use for grinding corn and other activities	William Armstrong builds the first hydroelectric scheme to provide lighting in his Northumberland home

Hydropower

The hydrological or water cycle is solar-powered. Evaporation takes up heat; without it, the average temperature of the oceans would be about 65 degrees Celsius. Some of that energy is given back as water condenses, powering towering cumulonimbus storm clouds and building up electrical charges, which discharge as thunder and lightning. More of the potential energy given to the water in taking it up to fall as rain on mountain ranges is returned as hydropower, in the erosion and transport of rocks and sediments, in the thunder of waterfalls and, occasionally, when tapped by humans to drive waterwheels or hydroelectric generators. There were nearly 3,000 TW hours of hydroelectricity generated in 2006: 20 per cent of the world's total electricity consumption – more than from nuclear power – and 88 per cent of renewable energy electricity generation.

Where the water goes Water is fully engaged in the dynamic processes on Earth, cycling between ocean, air and land. Every day, about 1,170 cubic kilometres (280 million cubic miles) of water evaporates, 90 per cent of it from the oceans. The average length of time water spends in each reservoir varies considerably. In deep aquifers and in the Antarctic ice cap, it is of the order of 10,000 years. It takes about 3,000 years to cycle through the oceans and a matter of months for rivers. But it spends on average a mere nine days in the atmosphere in each cycle.

Distillation The water cycle provides the Earth with a constant distillation and purification service. Evaporation from the ocean leaves salt and other contaminants behind. In the atmosphere, water dissolves carbon dioxide to make a weak acid, which promotes chemical weathering of

1881	**1928**	**1984**	**2008**
First hydroelectric power plant at Niagara Falls, USA	Hoover Dam becomes the world's biggest hydroelectric power station, rated at 1,345 MW	Itaipu in Brazil becomes the world's largest hydroelectric plant, generating 14,000 MW	Three Gorges Dam in China opens with a capacity of 22,500 MW

rocks, but the distillation provides water that is otherwise essentially pure to rivers, lakes and aquifers. Those aquifers add further filtration to the water and may add other dissolved minerals, which serve as nutrients for plants and aquatic organisms.

Water of life Almost all the chemical processes of life take place in aqueous solution, so liquid water is vital. In most plants and animals, water is by far the most abundant substance. As well as providing a chemical substrate, it plays a key part in one of life's most important reactions: photosynthesis. Using an ingenious system of enzymes and membranes, the chloroplasts in plant leaves are able to split the water molecule and combine it with carbon dioxide, producing oxygen and the hydrocarbons that build biomass. This was the chemical process that transformed the Earth's atmosphere more than 2 billion years ago and that keeps oxygen and carbon dioxide in delicate balance today.

Surviving drought So long as there is some water available, living organisms have developed sophisticated ways of coping with salinity, drought and freezing temperatures. All three challenges can have similar effects. Salty water can suck moisture out through cell membranes; drought can starve cells of water; and frost can turn water into ice, bursting cells or at least making water unavailable. Desert plants use a special sugar called trehalose to protect their dehydrated cells. Tiny invertebrates called tardigrades (they look like millimetre-long six-legged teddy bears) can withstand freezing in liquid nitrogen; they just get up and walk away when they thaw out. Fish in polar waters have a special antifreeze in their blood that suppresses the formation of ice crystals. In winter, the blood of the Antarctic cod gets so cold that it is way below freezing, and just touching it with an ice crystal will make it freeze solid!

> **For many of us, water simply flows from a faucet [tap], and we think little about it beyond this point of contact. We have lost a sense of respect for the wild river, for the complex workings of a wetland, for the intricate web of life that water supports.**

Sandra Postel, *Last Oasis: Facing Water Scarcity*, 1997

Eutrophication

One hazard of intensive agriculture is that excess chemical fertilizers, in particular those containing nitrogen and phosphorus, dissolve in groundwater and surface run-off, adding excessive nutrients to rivers, lakes and coastal waters. This, together with untreated sewage, can sometimes result in algal blooms that, fertilized by the nitrates and phosphates, use up all the dissolved oxygen in the water, leading to it becoming anaerobic or eutrophic. As a result, the water at the bottoms of some lakes and sheltered seas can no longer support life other than smelly anaerobic bacteria. Around half the lakes in Europe, Asia and the Americas are now eutrophic.

Many bacteria, in particular a species called *Pseudomonas syringae*, have proteins on their surface that encourage the formation of ice crystals. There is evidence that these play an important part in the atmosphere in the formation of ice clouds, which lead to rain and hail. Crops can be protected from frost damage by spraying on a genetically modified version, known as ice minus bacteria, though there are fears that widespread use of this may affect rainfall.

Water and climate Water vapour in the atmosphere is an important contributor to the greenhouse effect. Without it, average world temperatures would be about −30 degrees Celsius. There is evidence from climate models that global warming will lead to increased evaporation and, potentially, enhanced warming. It may also intensify the hydrological cycle, increasing rainfall in already wet equatorial and high latitudes, while worsening droughts in already arid areas in between.

the condensed idea
Water: bearer of energy, sustainer of life

30 Carbon cycle

If water is life's blood, carbon is its body. And, like water, carbon integrates life with a complex cycle that affects the whole planet. For eons, life has helped maintain the carbon cycle, keeping the insulating carbon dioxide blanket in the atmosphere in tune with increasing solar radiation. Now, human activity threatens to undo the work of the past.

The carbon cycle involves significant exchanges between huge reserves. The amount of carbon held in rock, soil, ocean, biomass and atmosphere is vast, and billions of tons cycle between them every year. But just a small imbalance in that cycle could make a big difference to our climate.

Carbon reserves The Earth's crust holds the fossilized remains of the early atmosphere. Two billion years ago, as life got going, it started to consume the insulating atmospheric blanket that kept it warm, converting the carbon dioxide ultimately into limestone and chalk, coal and oil. There are roughly 100 million billion tons of carbon in the Earth's crust, approximately the same as in the atmosphere of lifeless Venus. The implication is that, but for life, we would have a hothouse atmosphere similar to that of Venus.

The amounts of carbon held near the surface are smaller, though still very large. The greatest reserves are in soils and the deep ocean, followed by surface waters of the sea and the biosphere on land. Carbon dioxide in the atmosphere represents only a small fraction of the other reserves, at about 750 billion tons, of which around 200 billion tons are exchanged annually

timeline

1789	1800	1859	1896
Antoine Lavoisier shows respiration is the same process as combustion	Ice cores show atmospheric CO_2 level is 290 ppm	John Tyndall shows that carbon dioxide and water vapour absorb heat radiation	Svante Arrhenius suggests human CO_2 release might cause climate warming

Ocean carbon

The oceans contain 60 times as much carbon as the atmosphere and dissolve about 92 billion tons as carbon dioxide from the air each year. Some of that cycles round in the surface waters of the ocean, part of it being consumed by phytoplankton but getting released again. Ninety billion tons is returned to the atmosphere. Around 2 billion tons are removed from circulation and sink to the ocean floor. This is largely thanks to descending ocean currents, but also the hydrodynamic properties of copepod poo! These tiny planktonic animals feed on the microscopic algae and excrete faecal pellets that are large and dense enough to sink deep in the ocean as fine snow.

between the atmosphere and other reserves. Roughly half of that is gas exchange with the oceans; the rest is accounted for by photosynthesis and respiration on land.

Tipping the balance Those fluxes balance out more or less, with a slight surplus being removed from circulation through sedimentation in the ocean. But the figures exclude the 5.5 billion tons of carbon dioxide emitted into the atmosphere per year as a result of burning fossil fuels. A fraction of that excess, perhaps 10 per cent, seems to be getting removed from the atmosphere by processes in the ocean and probably by forests on land. But the rest is not.

In 1958, Charles Keeling started collecting air samples from the top of the Mauna Loa Observatory in Hawaii – far from any source of pollution. He measured carbon dioxide content and found a regular seasonal variation. But as he continued his measurements, every year the total CO_2 concentration got higher and higher. The resulting 'Keeling curve' has

1958	1988	2005	2009	2011
Charles Keeling begins measuring CO_2 in Hawaii	Intergovernmental Panel on Climate Change established at Rio climate conference	Kyoto treaty, signed by all industrialized countries except USA, comes into effect	Copenhagen conference fails to get binding agreement on CO_2 reduction	Carbon dioxide concentration reaches 391 ppm

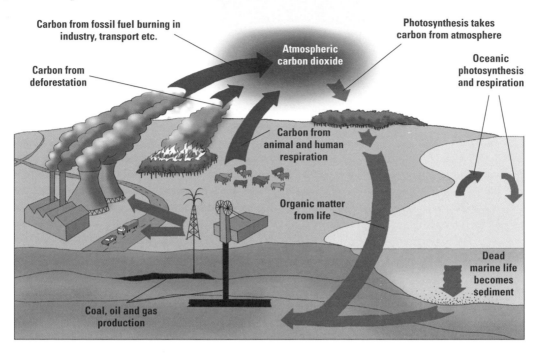

Carbon from fossil fuel burning in industry, transport etc.

Photosynthesis takes carbon from atmosphere

Atmospheric carbon dioxide

Oceanic photosynthesis and respiration

Carbon from deforestation

Carbon from animal and human respiration

Organic matter from life

Dead marine life becomes sediment

Coal, oil and gas production

The principal components of the carbon cycle.

become iconic. It starts in 1958 at 320 parts per million (ppm); by 2011 it had reached 391 ppm. At this rate, it will pass 400 ppm in 2015 and could reach anywhere between 450 and 850 ppm by the end of the century.

Greenhouse effect In 1859, working at the Royal Institution in London, John Tyndall was measuring the ability of different gases to absorb heat radiation. He tried nitrogen and oxygen, with little effect. Then he tested water vapour and carbon dioxide. The results were spectacular. What he described we now know as the greenhouse effect. Light from the Sun passes easily through the atmosphere and can be reflected just as easily back into space. But some of it warms the ground and is re-emitted as infrared or heat radiation. That gets absorbed by carbon dioxide, water vapour and other so-called greenhouse gases,

> **❛You will die but the carbon will not; its career does not end with you. It will return to the soil, and there a plant may take it up again in time, sending it once more on a cycle of plant and animal life.❜**
>
> **Jacob Bronowski**

Forest carbon

Plants and animals on land contain over 6 trillion tons of carbon. Forests contain 86 per cent of the carbon that is above ground, as well as a great deal in the soil. Around 100 billion tons of carbon is taken up by photosynthesis on land each year, of which about 60 per cent is retained in biomass such as timber. But it is quickly released again by deforestation through burning and through disturbing the soil. It is possible that forests will make a small contribution to removing excess carbon dioxide from the atmosphere, as plants tend to grow quicker in higher concentrations of carbon dioxide.

trapping the heat in the troposphere and warming the climate. Without it, our planet would have frozen over long ago, but as carbon dioxide levels rise, so do global temperatures. We will examine the possible consequences of that in the next section.

Methane Carbon dioxide is not the only form of carbon in the atmosphere. There are traces of methane, too. This is normally produced by bacterial activity – in wetlands, tundra, ocean sediments and the digestive tracts of cattle. Methane is also a greenhouse gas, and a more powerful one than carbon dioxide. Methane levels are rising due to intensive agriculture and the warming of Arctic tundra. But there is a far bigger potential source. Huge reserves of methane lie frozen on the ocean floor in gas hydrates, which could be released if the ocean warms or sea level falls. Fifty-five million years ago, a sudden release of carbon into the atmosphere was followed by rapid warming of the climate, and this is believed to be due to the release of huge quantities of methane from gas hydrates on the seafloor.

the condensed idea
A small imbalance in the carbon cycle can upset climate

31 Climate change

We are fortunate that, since Neolithic times, our species has enjoyed a relatively stable climate. But the geological record reveals that this was not the case in the past, and computer models of our climate system suggest it will not be the case in the future.

Working out if the climate is changing is harder than it may seem. We are very used to extremes of weather – heat waves, harsh winters, floods, droughts and so on – but that's not the same as climate. To identify climate change, you need to take many measurements worldwide over a long period and to uniform standards.

Taking the Earth's temperature Accurate temperature records measured with thermometers only go back about 150 years; before then, you need to use proxies. These can include historical accounts of harvest times and the extent of winter ice, as well as natural records such as the width of tree rings and isotope ratios in sediments and ice cores. Over the last century, these indications can be calibrated against accurate modern temperature records.

Medieval warm period All the records indicate that there was a warm period lasting from about AD 950 to 1250. Primary sources show that monasteries in England had flourishing vineyards at this time, while Viking settlers were farming successfully around the coast of Greenland.

timeline highs and lows from the past climate record

55 Ma	3.25 Ma	18 ka	12.8–11.6 ka	950–1250
Sudden warming, linked to methane release	Onset of rapid glacial cycles marking the last ice age	Last glacial maximum	The Younger Dryas: sudden cold spell	Medieval warm period. Vikings farming in Greenland

The little ice age Between about 1550 and 1850, and particularly in bursts beginning in 1650 and 1770, severe cold spells occurred. This is reflected in the 1565 winter paintings of Pieter Bruegel the Elder and in historical accounts including that of the frozen Baltic in 1658 and of frost fairs held on the frozen River Thames between 1607 and 1814. The period became known as the little ice age. These harsh winters are reflected in closely spaced growth rings on trees; it has even been suggested that the bitter cold produced the dense wood that helped Cremona violin makers such as Stradivarius to manufacture such resonant instruments.

Several explanations have been offered, but the most likely one is a dip in solar activity. The Sun follows an 11-year cycle of rising and falling activity, as reflected by sunspots. Space observations have confirmed that slightly more radiation, especially at ultraviolet wavelengths, comes from the Sun when there are more sunspots. Between 1645 and 1715, sunspot activity almost stopped – an interval known as the Maunder Minimum.

800,000 years of ice

When it snows, air fills the gaps between the snowflakes. In Greenland and Antarctica, where the snow never melts, it builds up in annual layers and slowly gets compressed. This turns it into ice, from which the air cannot escape. So ice contains a record of both the snow and the atmosphere. The deepest ice core ever drilled comes from Dome C in the middle of the eastern Antarctic ice sheet, providing a record going back 800,000 years. Oxygen isotope ratios in the water give clues to the surrounding ocean temperature, while the gas samples reveal carbon dioxide levels at the time. The two follow one another closely and reveal that the carbon dioxide level has not been higher than it is now for 800,000 years.

1645–1715	1816	1991	1998–2010
Maunder Minimum in solar output: little ice age with frost fairs on the Thames	The 'year without a summer' follows the eruption of Tambora	Aerosol from Mount Pinatubo cools world temperature by half a degree	Warmest years on record since 1850

Volcanic effects Among the events that leave their mark in the climate record are major volcanic eruptions. The best documented was that of Mount Pinatubo in the Philippines in 1991, which injected so much fine ash and sulphate aerosol into the stratosphere that there was a significant reduction in the amount of sunlight reaching the Earth's surface, causing a half a degree drop in global average temperatures over the next two years. The eruption of Tambora in Indonesia in 1815 was followed in Europe by what became known as the year without a summer. Crops failed and, during the winter of 1816–17, thousands starved or froze to death. An even larger eruption of Tola in Sumatra about 70,000 years ago may have almost wiped out the early human population.

Global dimming Volcanic aerosols in the atmosphere can form a faint haze that reflects sunlight and cools the temperature. So can aerosols from pollution. The result is global dimming. Systematic observations begun in the 1950s show a reduction of about 4 per cent in the amount of sunlight reaching the Earth's surface between 1950 and 1990. That, and the volcanic dust from Mount Pinatubo that followed, may have partially masked the effects of global warming due to greenhouse gases; the rapid warming in the late 1990s may, ironically, be partly due to reductions in pollution.

> ❛Climate change is the most severe problem that we are facing today, more serious even than the threat of terrorism.❜
>
> **Sir David King,**
> **UK Government**
> **Chief Scientific Adviser, 2004**

Ice ages Going back over the geological record, there seem to have been much bigger changes in climate than any experienced in human history. Reaching back through 3.25 million years takes us through a series of ice ages (see next chapter). The glacial maxima are closely matched by minima in atmospheric carbon dioxide, but rises in temperature seem to come a few hundred years before rises in CO_2. Climate change sceptics have used this evidence to argue that warming is not caused by CO_2. But in fact the lag is probably due to what is known as forcing, or positive feedback. Changes in the Earth's orbit lead to warming of the oceans which start releasing CO_2. That, in turn, causes further warming.

Oxygen isotopes

Most of the oxygen on Earth has an atomic weight of 16, but there is also a slightly heavier isotope, oxygen-18. Water containing the lighter oxygen-16 evaporates slightly more easily at lower temperatures and so the ratio of oxygen-18 to oxygen-16 reflects ocean surface temperature. Ocean surface oxygen left behind after evaporation becomes incorporated into the calcium carbonate shells of tiny foraminifera preserved in sediment cores. The higher the oxygen-18 ratio in them, the lower the surface temperature of the water where they developed. Conversely, the higher the oxygen-18 ratio is in ice cores, the warmer the ocean from which it evaporated. There are other complicating factors, but calculations can track ocean temperatures over hundreds of millions of years.

The geology of climate Going back further still, based on oxygen isotope ratios, we come to a world with much higher CO_2 levels and, sometimes, much higher temperatures. There are other ice ages, but in between them, global temperatures are typically as much as 10 or 15 degrees Celsius higher than today. The implication is that there are different stable states for our planet's climate, with quite a delicate balance between them. The question is: are we now heading towards a tipping point into a new and much warmer world?

the condensed idea
The climate keeps changing

32 Ice ages

The last 3.25 million years of Earth's history have been characterized by rapidly alternating periods of glaciation and warm interglacials. The cold spells are known as ice ages and are caused by wobbles in the Earth's orbit, leading to changes in the solar radiation on Earth. They may have driven key stages in human evolution.

Erratics In the 18th century, various naturalists noticed large, so-called 'erratic' boulders in Alpine valleys and suggested that they were deposited by glaciers that were once more extensive than they are now. In 1840, after a decade studying fossil fish, a bright young Swiss geologist named Louis Agassiz turned his attention to the vast superficial deposits of sand, gravel and boulders surrounding the Alps. He concluded that they had been dumped there not by individual glaciers, but by a vast ice sheet that must once have covered the entire mountain range. It represented, he proposed, an ice age.

The extent of ice Agassiz speculated that the ice extended from the North Pole right the way to the Mediterranean, as well as across the Atlantic and the whole of North America. We now know that the Alpine ice cap remained isolated, but the polar ice sheet was indeed vast, extending across Scandinavia and almost as far as the Thames in Europe and south of the Great Lakes in North America. The ice was up to 3,000 metres (9,843 ft) thick and held so much water that global sea level was reduced by about 110 metres (361 ft), turning the continental shelves into dry land and creating land bridges for migrating animals and humans.

timeline

2.4–2.1 Ga	850–630 Ma	460–420 Ma	360–260 Ma
Earliest known glaciation: the Huronian	Cryogenian period: the most severe glaciation the world has known	Large polar ice caps develop, though glaciation is not as severe as the Cryogenian	Karoo glaciation, leaving glacial deposits in South Africa and Argentina

MILUTIN MILANKOVICH 1879–1958

Milankovich was a Serbian mathematician and civil engineer who was made a professor at the University of Belgrade in 1909. He began calculating the amount of solar radiation falling on the Earth and how it varied and, in 1914, published his first paper on an astronomical theory of the ice ages. He moved to the Austro-Hungarian Empire to marry, but when the First World War broke out he was interned as a Serbian citizen and spent the next four years in confinement working on his theories. In 1920, he published a monograph detailing how the complex orbital variations interact to produce the Milankovich cycles that initiate ice ages. He went on to calculate the surface temperature of Mars, showing that it could not support life. By 1941, he had completed a book summarizing his scientific work. It had just been printed when war broke out again and the printer in Belgrade was bombed, leaving Milankovich with the only surviving copy.

Advance and retreat Some outdated textbooks still suggest that there were four glacial episodes during the last ice age. In fact, the story is much more complex, with at least 20 episodes now identified, interspersed with quite warm interglacial periods during which lush vegetation and extensive fauna returned. Early human hunters followed the migrating herds across the plains of what is now the North Sea, and there is evidence that Britain was occupied on six or seven separate occasions, first by *Homo heidelbergensis*, then by Neanderthals and finally by *Homo sapiens*.

Causes The precise cause of the ice ages is not clear, but it is likely to be a complex interaction of several factors, the clearest of which seems to be variations in the Earth's orbit. These are known as Croll–Milankovich

3.25 Ma	2.4 Ma	800 ka	18 ka
Beginning of present ice age sequence	Ice age intensifies	Base of the oldest ice core, which records eight glacial maxima since this date	End of the last glacial maximum

cycles, after the Scottish scientist who proposed them, and the Serbian engineer and mathematician who developed the idea. They bring together three factors: the eccentricity of the Earth's orbit around the Sun; the tilt of the Earth's rotation axis; and the precession of that axis – in the same way that the axis of a spinning top draws out a circle. These all vary with different periods – around 400,000 years, 41,000 years and 26,000 years, respectively. The result is that different amounts of solar radiation fall on different regions of the Earth in different seasons. According to this cycle and without any human-induced global warming, we might be due for another glacial period in 15,000 years.

Problems This 'solar forcing' seems to fit in with the glacial cycles, but it cannot be the complete explanation. For one thing, for the first 2 million years, the cycles seem to follow a 41,000-year period; but for the last million years, this changes to a 100,000-year cycle. That may be something to do with the time lag of the ice behind the solar forcing. Also, the variations in solar radiation seem much smaller than the climate variations that result. This may be due in part to various positive feedback mechanisms. For example, the albedo – the amount of solar radiation our planet reflects back into space. As the white ice spreads, more sunshine is reflected and so cooling increases. When warming starts, more CO_2 is

Ice age deserts

Many of the world's deserts are concentrated around 30 degrees north and south of the equator, where cool, dry air descends from the Hadley cell of atmospheric circulation. It is clear from windblown deposits on land and in ocean sediment cores that these deserts were more extensive and that North African lakes were at their driest during the Last Glacial Maximum, about 18,000 years ago. This is believed to be due to a weakening of the African and Asian monsoons as a result of smaller differences between the temperature of land and sea. The drying-out of East Africa at the onset of a glacial cycle may have been a driving force in human evolution.

released from the oceans and so warming increases. The interactions are complex, but the Milankovich cycles seem to be central.

The initial trigger for the last ice age may have been the closing of the Isthmus of Panama, separating the circulation in the Atlantic and Pacific oceans.

Rebound The weight of all that ice pressed down on the northern continents and pushed them into the mantle. When the ice melted, they started to rebound – like a cork bobbing to the surface, only much, much slower. Such is the stiffness of the mantle that they rise at only about 1 centimetre (½ in) per year – a process that is still continuing 10,000 years after the last ice age. The results include raised beaches covered with seashells in Scotland, 80 metres (262 ft) above the present sea level.

Ancient ice ages The ice ages of the last few million years are not the only ones in Earth's history. There appear to have been at least five major episodes of glaciation, interspersed with warm spells during which our planet had no polar ice caps at all. The first probably began about 2.4 billion years ago and may be linked to the rise of photosynthesis, with marine algae using up carbon dioxide from the atmosphere. It is marked by rocks around Lake Huron in Canada, containing drop stones – large stones carried on sea ice and released in deep water. Next came the most severe ice age, in the late Precambrian, which we will return to in a few pages time. Another occurred at the end of the Ordovician, and a fourth lasted 100 million years, starting 360 million years ago and leaving traces in South Africa and Argentina (see timeline).

> **The glacier was God's great plough ... set at work ages ago to grind, furrow, and knead over, as it were, the surface of the Earth.**
>
> **Louis Agassiz, 1807–73**

the condensed idea
Small changes lead to a big freeze

33 Ice caps

The 'ends' of the Earth are capped with ice. They are places of great beauty and scientific interest. For much of the geological past, the Earth did not have polar ice caps and sea levels were correspondingly higher. The present ice caps have persisted for millions of years. But could they now be under threat? The climate is warming faster around the edges of polar regions than anywhere else on Earth.

Frozen sea The North and South Poles are very different places. The North Pole is in the middle of ocean surrounded by continent. The South Pole is on a continent surrounded by ocean. As a result, and with the exception of the massive Greenland ice cap, the northern ice is floating in the sea, which means that, especially in summer, it can easily start to melt and break up and is forever on the move. On the other hand, if it does melt, floating ice will not change the sea level.

Frozen continent Antarctica is bigger than the USA and almost entirely covered by ice. In all, there are nearly 14 million square kilometres (5,405,430 square miles) of ice; the ice is on average nearly 2 kilometres (1¼ miles) thick and makes up 75 per cent of the fresh water on Earth. Because of the ice thickness, it is the highest continent as well as the coldest, the driest and the windiest. The coldest temperature ever recorded on the surface of the Earth was −89 degrees Celsius at the Russian Vostok Station on East Antarctica.

timeline milestones of polar exploration

1820	1841	1845–48	1903–06	1909
The Russian Captain Bellingshausen and his crew are the first to sight Antarctica	James Clark Ross reaches the ice shelf that is named after him	Sir John Franklin's ill-fated expedition to the Northwest Passage	Roald Amundsen negotiates the Northwest Passage	Robert Peary is possibly the first to reach the North Pole (disputed)

Sea ice Ice develops at both poles because they receive so little of the Sun's radiation. North of the Arctic Circle and south of the Antarctic Circle, the Sun does not rise for several months during winter, and even in summer it never reaches a high angle. Ice caps on land have all formed from accumulating snow. Sea ice begins to form when the sea water itself freezes, though it can subsequently accumulate snow on top. Most of the sea ice around Antarctica is seasonal and seldom more than six months old and a couple of metres thick. Arctic sea ice can persist for several years, reaching thicknesses of 4 or 5 metres (13–16 ft) or more where it is compressed into ridges.

Northwest Passage From the late 15th century, the search for a 'northwest passage' to the Pacific became an obsession among some navigator explorers. Many unsuccessful and sometimes fatal expeditions were made over the next 400 years, including that of Sir John Franklin in

Hidden lakes

Russia's Vostok stock research station – the coldest place on Earth – is on an unusually flat area of ice in Eastern Antarctica. Radar and seismic surveys have shown that this is because it is on a lake. The ice is nearly 4 kilometres (2½ miles) thick, but beneath it is a lake of liquid water. The size of Lake Ontario and, at over 300 metres (984 ft) deep, three times the volume, it is the largest of around 140 sub-glacial lakes in Antarctica.

It is possible that it has been isolated for millions of years and may contain strange life forms nourished by hydrothermal springs in the lake bed. At the time of writing, a Russian ice core has come within 50 metres (164 ft) of the water and may sample it in the next season. Meanwhile, British scientists are planning to drill another sub-glacial lake, Lake Ellsworth, which is the size of Lake Windermere.

14 December 1911	**17 January 1912**	**1914–17**
Roald Amundsen leads the first party to reach the South Pole	Robert Falcon Scott reaches the South Pole	Ernest Shackleton's Antarctic expedition on *Endurance* – last great voyage of polar exploration

> **❝Had we lived, I should have had a tale to tell of the hardihood, endurance and courage of my companions which would have stirred the heart of every Englishman.❞**
>
> **Robert Falcon Scott, final words in his diary, 25 March 1912**

1845. It was the Norwegian Roald Amundsen who finally made it through the ice, but it took him from 1903 to 1906 to do so. Today, in many summers, it is easy to pass through. Every year more of the Antarctic sea ice seems to break up and melt, and every September there is more open water. Satellite measurements show that the ice that remains is getting thinner too. Climate models suggest that summer sea ice will have completely disappeared from the Arctic by the end of the century, though it may be all gone by 2050 at its present rate of retreat.

As Arctic waters become more accessible, the Northwest Passage could become a major trade route and the seafloor might be exploited for oil and minerals. As the waters warm up, large deposits of methane hydrate could become unstable, leading to significant releases of this greenhouse gas. As sea ice forms, it expels salt, producing freshwater ice. As a result, the remaining water becomes more salty and dense; that helps to produce the Atlantic bottom water that maintains the conveyor belt of ocean currents. If less ice forms, it might upset ocean circulation.

The isolation of Antarctica Once, 170 million years ago, Antarctica was part of Gondwanaland, a tropical supercontinent with forests and dinosaurs. The break-up began with Africa splitting away 160 million years ago, followed by India 125 million years ago and Australia and New Zealand 40 million years ago. Ice began to form on the cooling continent, but it was not until the Drake Passage opened between Antarctica and South America less than 34 million years ago that it was fully gripped by ice. That enabled an ocean current to circulate from west to east right around Antarctica, isolating it from incoming warmth.

Ice on the move Ice never stays still for long. In the centre of Eastern Antarctica, where it is 3,000 metres (9,843 ft) thick, ice accumulates over tens of thousands of years, but at the same time,

Lost landscape

Ice is transparent to airborne radar; this resulted in some unfortunate ultimate errors for early aviators. In recent years, research planes have been surveying the landscape deep beneath the ice of Eastern Antarctica. They have revealed a spectacular landscape of mountain ranges and fjords – a record showing how the eastern Antarctic ice sheet advanced and retreated over the last 34 million years.

ever so slowly, it flows. Around the edges and in glaciers it flows more quickly, spreading out over the sea to form floating ice shelves hundreds of metres thick and eventually breaking up into vast icebergs. Its motion on land is lubricated. Pressure from above and heat from below cause melting and the ice streams ride on a slippery layer of wet mud. As long as the outward flow of ice is matched by accumulating fresh snow, all is well.

Antarctica's soft underbelly Western Antarctica is different. Most of the land is below sea level, though the ice towers far above. But that makes the region especially vulnerable to the warming ocean currents that encircle it. The Pine Island Glacier is the size of Texas. It is the biggest Antarctic glacier and, in the last few years, it has started accelerating and thinning at an alarming rate. An unmanned submarine that ventured under the lip of the glacier revealed that it has melted clear of a ridge of rock that was preventing it from collapsing into the sea. If the whole glacier were to collapse, it would raise sea level worldwide by a quarter of a metre (9 in). The adjacent group of glaciers would raise sea level by 1.5 metres (5 ft) if they melted.

the condensed idea
Ancient lands of ice beginning to melt

34 Snowball Earth

There have been five great ice ages in the Earth's history, but none was more severe than that of the Cryogenian period in the late Precambrian. For two episodes of up to 20 million years, ice reached into the tropics. One of the most fascinating geological controversies of our time surrounds whether the whole planet froze solid to form a Snowball Earth and if so, how it ended.

Dropstones For nearly a century, geologists have been noticing glacial deposits in surprising places – some of them far from present polar regions. But the record of late-Precambrian rocks is patchy, and it is sometimes hard to recognize glacial deposits in strata that old. One feature for which there is no other explanation is the presence of dropstones: large boulders deposited far from land in otherwise fine-grained marine sediments. There is only one known mechanism to get them there: on rafts of floating ice from glaciers.

Precambrian continents With the development of plate tectonics in the 1960s came the realization that the continents have not always been in their present positions. The orientation of magnetic particles in rocks can reveal their latitude at the time they were deposited. That showed that glacial deposits around 640 million years old in Canada, Greenland, Svalbard, Namibia and Australia were all near the equator – yet they all carry evidence of glaciation.

timeline discovering the Snowball

1949	1964	1966	1992
Douglas Mawson shows that Precambrian glacial deposits are widespread on all continents	Brian Harland reports dropstones from Svalbard and Greenland, then tropical latitudes	Mikhail Budyko calculates that, if ice extends to a latitude of 30 degrees, it will continue to the equator	Joe Kirschvink coins the term Snowball Earth

Positive feedback The fear of global nuclear war in the 1960s led to calculations about the cooling effects of the resulting clouds of dust and aerosol. The numbers showed that, if ice extended as far as 30 degrees north and south, it would increase the amount of sunlight reflected back into space so much that it would lead to positive feedback: the more the ice grew, the colder it would get. This would lead to a global freeze. The process might be triggered by variations in the Earth's orbit, a reduction in solar output and perhaps an increase in cloud triggered by cosmic rays as the solar system passed through a spiral arm of the galaxy.

Eternal snowball In a 1992 paper, Joe Kirschvink of Caltech coined the term 'Snowball Earth' to describe such a freeze and suggested that it might have happened in the Precambrian. But critics raised a problem. With the planet a bright, white snowball, and no oceans to evaporate and create cloud, solar radiation would continue to be reflected into space and the cycle could never be broken.

PAUL HOFFMAN b. 1941

Canadian geologist Paul Hoffman is a determined man. He ran in several marathons before coming ninth in the 1964 Boston Marathon. But he realized that he was unlikely to be able to break the world record or win an Olympic medal. Only the best was good enough and so he opted for geology, becoming a professor at Harvard – one of the most prestigious positions in the world. Every year he went on field trips to remote parts of the world and in particular the hills of northern Namibia, where he studied the Precambrian deposits, proving that they were glacial and suggesting that the overlying carbonate deposits came after a sudden thaw. He has been an ardent advocate of Snowball Earth ever since.

1998
Hoffman and Schrag publish key paper on the glacial deposits of Namibia and overlying carbonates

2006
Snowball Earth conference questions the global extent of glaciation

2010
Glacial deposits in Canada accurately dated to 716.5 million years, when the region was equatorial

> **The problem with the snowball is that this should have been the greatest environmental calamity of all time and yet we can't find the bodies.**
>
> **Prof. Guy Narbonne,**
> **BBC TV *Horizon*, 2001**

Kirschvink suggested a solution. One heat source that continues regardless of ice is that of the Earth's interior. Volcanoes would continue to erupt, releasing carbon dioxide. Without open water, the gas could not be dissolved in the ocean and it would build up over 10 million years to a point where the atmosphere was 10 per cent carbon dioxide. Even global ice could not survive in that greenhouse: there would be rapid melting and a spectacular heat wave.

Cap carbonates Further evidence came in a 1998 paper by Harvard geologist Paul Hoffman. He had studied the glacial deposits of Namibia and noticed that they were often capped with limestone. These so-called cap carbonates, he argued, were the result of rapid chemical weathering of rocks in the first warm rain to fall in 10 million years, drawing down the CO_2 and depositing limestone.

There are naturally two stable isotopes of carbon: carbon-12 and carbon-13. Living organisms tend to concentrate carbon-12, but the cap carbonates are not depleted in carbon-13, suggesting they came from CO_2 with a volcanic origin. Also, at the base of the cap carbonates is a layer enriched in iridium. That is rare on Earth but abundant in meteorites and dust from space. Ten million years of ice would accumulate iridium-rich dust.

Anoxic waters Glacial periods are marked by layers of banded iron formation, and that is precipitated by the oxidation of soluble ferrous oxide into insoluble ferric oxide. For the ferrous salts to accumulate you need large bodies of anoxic water, such as might form if the oceans were isolated by ice.

Reaching a verdict The Snowball Earth theory makes a fascinating story, supported by compelling evidence, but the scientific jury has yet to reach a unanimous verdict. For one thing, how could primitive life have survived such a catastrophe? Life at that time was all in the sea and consisted mainly of algae and bacteria. How could organisms get the

The Snowball and life

By 720 million years ago, the land was still bare but there was abundant life in the sea. Photosynthetic algae and cyanobacteria had already transformed the atmosphere, drawing down carbon dioxide and releasing oxygen. Snowball Earth must have been a terrible calamity from which life scarcely survived. But survive it did and made good use of the opportunity. Once the ice melted, suddenly there was a new frontier of nutrient-rich waters with no competition. Shortly after the end of the last glaciation, we find the first widespread evidence for a diversity of multicellular animal life. As we will discover in the following sections, the rest is palaeontology.

light they needed for photosynthesis through thick ice? A possible answer is that, if freezing is slow, ice can be almost transparent. Photosynthesis continues in Antarctica's dry valleys under 5 metres (16 ft) of ice.

The big question is just how complete the freeze was. Dropstones could have been rafted to the equator by icebergs in open water, and, critics say, even a small area of open water would be enough to dissolve atmospheric carbon dioxide, preventing its accumulation in the atmosphere. Another big unknown is the state of the Earth's magnetic field at the time. If it was not aligned close to the Earth's rotation axis, some of the glacial deposits may not have been as close to the equator as they seem. This and other criticisms have led to the proposal of a slushball Earth rather than a snowball, with areas of open water maintained at least seasonally.

the condensed idea
Did the entire Earth freeze over?

35 Deep time

If there is one idea in earth sciences on which all others depend, it is the concept of deep time. Without it, solid rocks cannot flow, mountains cannot rise, raindrops cannot wear rocks down and life cannot evolve by gradual mutation. And yet, to beings more familiar with time measured in hours, days and years, deep time is one of the hardest ideas to comprehend.

The perception of time We build up our perception of the physical scale of the Earth by direct experience. We can go for a long walk, admire a wide view, fly halfway around the Earth or study a photograph of our planet taken from space and thus build an understanding of the physical size of our planet from our own observations. Time is different. Most of the things we do break down into actions that last merely seconds. Our lives are governed by the passing of hours and days and we celebrate anniversaries on a yearly basis. But we can only just remember a few experiences from our early childhood and we have only second-hand reports of time before that. Imagining the time dimensions of our planet on that basis is like trying to comprehend the planet's depth when your concept of depth is based on the thickness of a piece of paper.

> **Geology gives us a key to the patience of God.**
>
> **Josiah Gilbert Holland, c.1870**

Generation upon generation A generation is about as far back as it is easy to comprehend. If we take a generation to be 25 years, then our great-great-great-grandparents lived in the 19th century at the time of Queen Victoria, five generations ago. Just 17 generations takes us back to the Spanish Armada and less than 40 to the Norman Conquest in 1066.

timeline geological time as one day on a 24-hour clock

0:00	02:00	06:00	10:00
Accreting dust and rock forms the Earth	Heavy bombardment. Oldest rocks	First clear fossil evidence for life	First free oxygen in the atmosphere

The cooling Earth

A century before Darwin proposed his theory of evolution, Georges-Louis, Comte de Buffon, suggested that animals change over time. He realized that this process called for a considerable length of time and performed experiments to see how quickly iron balls cooled from white heat to room temperature. Extrapolating that to something the size of the Earth, he concluded that our planet must be 74,832 years old. The idea was taken up by Lord Kelvin in the late 19th century, using more precise estimates of the cooling rate of molten magma. He decided that the Earth must be somewhere between 20 and 400 million years old, though in 1897 revised this to no more than 40 million years. But he did not know about the heat of radioactive decay.

It takes about 180 generations to reach back to the builders of Stonehenge, and even that is almost an afterthought in the history of the Earth. Four thousand generations ago, our ancestors were migrating out of Africa, but if, to use a different analogy, the lifetime of the Earth was expressed as a day on a 24-hour clock, that migration would have taken place just two seconds ago!

The truth dawns So it is perhaps not surprising that pioneering geologists were slow to recognize the true depth of time. They were probably also hampered by their own beliefs, and those of theologians, that the Earth was created in six days and that humans turned up fully formed on day six.

By the 18th century, it had become rare in Europe to be burned at the stake for religious heresy; the age of reason was dawning. Philosophers

20:30	21:00	22:00	23:38	23:58:40	23:59:58
Possible Snowball Earth	Cambrian explosion of marine life	First animals on land	Extinction of the dinosaurs	First hominins to walk upright	Fully modern humans migrate out of Africa

Darwin's need for time

One of the problems that most vexed Charles Darwin in formulating his theory of evolution was the immense amount of time he thought would be necessary for random change and natural selection. He worked out his own estimate for the length of geological time from looking at the Weald of Kent. It was clear that this represented a huge dome of rock that must once have been capped by chalk. Now the chalk remains in the North and South Downs, with older rocks in between. Using a guess at the thickness and a rather random estimate of the rate of erosion, he concluded that the process must have taken 300 million years. But he realized that this was little better than a guess and removed the figure from later editions of his book.

and scientists began speculating openly about the true depth of time and observing the speed of processes that might give them some clues. In France, Benoît de Maillet (1656–1738) had observed fossil shells high above sea level. He estimated the rate at which French harbours were silting up, which he assumed to be due to the sea level falling. His estimate of 0.75 millimetre per year meant that fossils in the highest mountains would have to be 2.4 billion years old.

In 1821, the Rev. William Buckland discovered hundreds of bones in a cave in Yorkshire. They came from hyenas and their prey, and even elephants and rhinoceros. He concluded that it was unlikely even the biblical flood would have washed so many bones up from Africa and that they must represent many more generations than the ten described in the Bible between Adam and Noah. Therefore the biblical account of the flood could not be confirmed by geology.

The father of deep time In 1788, James Hutton published his famous *Theory of the Earth* in which he laid down the principle of gradualism – that processes at work today can eventually accomplish all

> **We cannot take one step in geology without drawing upon the fathomless stores of by-gone time.**
>
> Adam Sedgwick, letter to William Wordsworth, 1842

the changes in the geological past (see chapters 24 and 25). That called for immense amounts of time to have passed and did not prove popular with theologians, but it set scientists thinking about the phenomenon of deep time.

In 1841, Charles Lyell, the Scottish geologist and an advocate of Hutton's work, visited Niagara Falls. There, the lip of the waterfall is set far back in a long gorge. Lyell found an old man who claimed that he could remember the waterfall being 45 metres (148 ft) nearer 40 years previously. On that basis (and allowing a margin for exaggeration), Lyell estimated that the 11-kilometre (7 mile) gorge had been carved out over 35,000 years. We now know that that is not long in geological terms, but it helped thinkers to go beyond the biblical timescale; it also helped Charles Darwin when he was thinking about the length of time needed for evolutionary change.

The fourth dimension in perspective We now know, from radio dating techniques (see chapter 4), that the Earth is 4.56 billion years old and the universe three times that age. It is still hard to comprehend that immensity of time, but it makes sense of a whole range of geological processes. Mantle convection, continental drift, mountain uplift and erosion all progress at typical rates of a couple of centimetres a year. On human timescales, that seems trivial. But over deep time, the mantle circulates like a cauldron of hot soup, continents waltz about the globe and mountain ranges rise and fall like the chest of a sleeping dragon. And entire species of plants and animals emerge, evolve and fall extinct.

the condensed idea
With time comes change

36 Stratigraphy

Sedimentary rocks are layered, and it seems almost self-evident to us that those layers must have built up one after another through successive periods of geological time. But the first people to realize this were to transform geology into an accurate science and develop maps that changed the world.

By the time he turned to geology, the young Danish scientist Nicolas Steno had moved from Copenhagen to study in Florence under the patronage of the Medici family. He had found fossils in the Tuscan hills and recognized that they recorded past life. But how did they get into the solid rock? Steno observed that sedimentary rocks were found in layers and deduced, correctly, that in order to include fossils, they must have been laid down in sequence under a liquid.

Basic principles Steno went on to lay down four fundamental principles in 1669. The principle of superposition stated that rocks are formed in sequence in layers, with the oldest at the bottom. The principle of original horizontality suggested that they were formed originally in flat, horizontal layers. The principle of lateral continuity stated that these layers once extended continuously across the Earth, except where something got in the way. And the principle of crosscutting discontinuity stated that anything that cuts through those layers must be younger than they are.

In sequence Though not accurate in every way, Steno's four principles are a good guide to the science of stratigraphy. Sedimentary rocks do indeed form in layers with the youngest at the top – though they can

timeline start dates of periods in the geological column (Ma)

542	488	444	416	360	300	254	200	145
Cambrian	Ordovician	Silurian	Devonian	Carboniferous	Permian	Triassic	Jurassic	Cretaceous

be folded so much that the sequence is occasionally reversed. With a few exceptions such as cross-bedding, layers were originally horizontal; and, though no single layer is global, finding the same sequence of layers in two separate places is a good indication that they are contemporary with one another.

Somerset canal It was over a century before these principles were developed further and put into practice. In the 1790s, William Smith was surveying the proposed route for a canal in Somerset. He needed to predict what sort of rocks would be encountered as the canal was dug and if they would be able to keep the water in the canal. He soon realized that the same sequence of rocks appeared in the same order each time he found them. Furthermore, he recognized that each layer had characteristic fossils in it that would help to identify it elsewhere. He became known as William 'Strata' Smith.

> **Organised Fossils are to the naturalist as coins to the antiquary; they are the antiquities of the Earth; and very distinctly show its gradual regular formation, with the various changes, inhabitants in the watery element.**
>
> **William Smith,** *Stratigraphical System of Organized Fossils,* 1817

NICOLAS STENO 1638–86

Niels Stenson was born in Copenhagen and later Latinized his name to Nicolas Steno. At the age of 21, he resolved not to accept things he read in books but to trust his own observation – a scientific principle ahead of his time. At first he studied anatomy, but when he was sent the head of a giant shark to study in 1666, he recognized that its teeth were almost identical to fossils he had found in rocks. He concluded that fossils were the remains of past life. While speculating as to how a fossil could get into a solid rock, he developed his principles of stratigraphy. He was born a Lutheran but, never accepting what he was told, found Catholicism more in tune with his observations and converted, eventually becoming a bishop.

65	56	34	23	5.33	2.59	11,450 years
Paleocene	Eocene	Oligocene	Miocene	Pliocene	Pleistocene	Holocene

WILLIAM SMITH 1769–39

As the son of an Oxfordshire blacksmith, and unlike many gentlemen scientists of his day, William Smith did not have a private income and so worked as a surveyor for Somerset landowners and the Somersetshire Coal Canal Company. That took him into the countryside on a daily basis, where he recorded the rock layers and the fossils within them. He soon recognized that both rocks and fossils appeared in a regular sequence that was repeated wherever those rocks were found.

In 1799, he produced a geological map of the Bath area and, having been dismissed by his employers, went on to produce his famous geological map of England, Wales and part of Scotland. He tried to make a living selling copies of the map, but it was plagiarized and he was undercut, forced into bankruptcy and imprisoned as a debtor. Only later in life was his contribution to geology recognized: he was awarded the Geological Society's first Wollaston medal in 1831.

If you looked around the world today, you would soon see that present-day deposits are not identical. A riverbed may contain gravel, while nearby salt flats are building up mud at the same time, and a sheltered sea is depositing limestone. Smith realized this but also recognized that similar fossils found in only a short sequence of layers could be used to show that the different sediments were contemporaneous.

The first geological map William Smith also noticed that the strata he was studying were dipping gently to the east. If Steno's principle of original horizontality held true, they must have been tilted by subsequent ground motion. The understanding that, in travelling from west to east across England, he was encountering progressively younger rocks helped William Smith to draw up his famous geological map, which he hand-coloured to show each major division of geological layers.

Naming the layers Many of the principal layers were given names. For some, the local names used by quarrymen were adopted. Other names were given by geologists and were either descriptive or reflected the localities where the rocks were found. It was a natural progression to try to group together sequences of broadly similar strata and this led to the naming of periods of geological time. One of the first to do this was the

German geologist Abraham Werner (1749–1817), who had coined the term neptunism. While he was wrong to think that granite was deposited underwater, he was quite right that the majority of sedimentary rocks had formed in that way. He divided them into primitive, transitional, secondary and tertiary. His Tertiary period still survives in the geological column today.

The names of other periods reflect the parts of the world where they were first identified. Cambrian, Ordovician and Silurian take their names from Welsh tribes. Devonian rocks are found in the county of Devon in south-west England. Jurassic rocks are found in the Jura mountains, north of the Alps. The Cretaceous comes from the Latin for chalk.

One of the great pioneers of stratigraphy was Sir Roderick Impey Murchison (1792–1871). As well as doing important work in southern England, the Alps and Scotland, his major contribution was the establishment of the Silurian period. A forceful character, he argued strongly for the divisions that he had proposed. This led him into a major disagreement with Sir Henry De La Beche, founder of the Geological Survey of Great Britain. De La Beche had found fossils normally associated with the Carboniferous coal measures in rocks he thought were Silurian and therefore tried to suggest that fossils could not be used to correlate strata. Murchison was able to prove that the fossils were in the base of the Carboniferous and that between there and the Silurian was an eroded layer, elsewhere represented by the old red sandstone deposits of south-west England. Thus the Devonian period was created to fill that gap.

Dating the layers In 1977, the International Commission on Stratigraphy was founded to try to pin down the divisions between the periods precisely and to give them absolute dates. The painstaking radiometric dating conducted in the first half of the 20th century by Arthur Holmes (see chapter 4) has been found to be remarkably accurate and has only been improved on slightly since.

the condensed idea
Layers of time

37 Life's origins

Life seems a magical, miraculous thing. It is no wonder that people thought it must be a divine creation. The origins of life are still among the great unknowns of science. But ancient traces and modern experiments are beginning to close in on life's secret.

Life essentials 'Simple' is not a term you could fairly apply to even the most primitive bacterium on Earth. All extant forms of life today have such subtle complexity that it is hard to imagine them originating by random chance. So what are life's essentials?

The heart of it would seem to be an ability to make copies or reproduce, and for that there needs to be something to copy: some code or set of instructions that define the nature of the organism. In all life on Earth that we know of that is the genetic code, carried by the double helix of DNA or, in a few cases, the single spiral of RNA. Then you need a mechanism to copy that: in the case of DNA, that is a complex system of proteins, enzymes and cellular structures. And, if evolution is to happen through random change, the copying mechanism should be not quite perfect, introducing the possibility of change. To perform those functions, living organisms need to extract energy from their environment – perhaps chemical or solar. And to contain all this complexity, they need some sort of membrane or cell wall.

The building blocks of life Picking up from Charles Darwin's idea of 'a warm little pond', in 1953 Stanley Miller conducted a now famous experiment, showing how many of the chemical building blocks

timeline

5th century BC	1800s	1861	1920s
Anaxagoras proposes the idea of panspermia	Most people still believe that life begins through spontaneous generation	Louis Pasteur proves in his sterile flask experiment that life is not spontaneously generated	Oparin in Russia and Haldane in England suggest first biochemical model for origin of life

of life could be made by passing electrical discharges through common gases (see box: Stanley Miller). But perhaps making the ingredients was not the problem. Carbonaceous meteorites, which must have fallen more frequently early in the Earth's history, have been found to contain the necessary amino acids and bases. The building blocks of life may have been abundant, but their presence still does not generate life, any more than random explosions in a scrapyard will produce a functioning motorcar.

Natural scaffolding What held all the chemicals together before there were cellular structures to do the job? One potential candidate is clay minerals. They can form in thin sheets and occasional defects in the crystal lattice might be reproduced in adjacent sheets. Another possibility is iron pyrites, or fool's gold, which was probably abundant in the absence of atmospheric oxygen.

STANLEY MILLER 1930–2007

Stanley Miller was researching his PhD at Chicago University under Harold Urey when he performed his most famous experiment. He took a flask containing a little water and the gases then thought to make up the early Earth's atmosphere and subjected it to simulated lightning in the form of electrical sparks over a period of several days. A dark-brown liquid accumulated in the bottom of the flask which was revealed on analysis to contain amino acids and other chemical building blocks of life. We now know that Miller got the composition of the atmosphere wrong: it would have been mostly carbon dioxide and nitrogen, not the reactive mixture of hydrogen, methane and ammonia that he used. But he did show that it is relatively easy to make complex organic chemicals.

1953	1986	1992	1996	2011
Stanley Miller creates chemical building blocks of life in the laboratory	Walter Gilbert at Harvard coins the term 'RNA world'	William Schopf reports 3.5 billion-year-old micro-fossils from Australia	David McKay claims to find micro-fossils in a Martian meteorite	Martin Brazier finds sulphur catalysing bacteria fossils in 3.4 billion-year-old rocks

One of the best candidates for the chemical basis of the first life is RNA. Like its more stable, double-strand cousin DNA, RNA can carry a genetic code. Crucially, it can also act as an enzyme, even catalysing its own reproduction.

Places to hide A 'warm little pond' would not have been a very safe place on the early Earth. Constant bombardment with radiation from space and with meteorites and asteroids would, some argue, have rendered the surface uninhabitable. Perhaps, they suggest, life got started around volcanic vents on the ocean floor, or even in hydrothermal systems underground.

Shadow biosphere As Darwin suggested, as soon as life got going it would have consumed all the available organic ingredients. But natural processes are not normally unique, so it is possible that life began many times and perhaps in different forms. Complex molecules can often exist in mirror-image forms. Life as we know it only makes use of left-handed molecules. Could the right hand of creation – evidence of a second genesis – still be surviving in isolated pockets on or within the Earth? One place to look might be in those hydrothermal systems that are too hot to support normal life.

> **It is often said that all the conditions for the first production of a living organism are now present, which could ever have been present. But if (and oh what a big if) we could conceive in some warm little pond with all sorts of ammonia and phosphoric salts, – light, heat, electricity &c. present, that a protein compound was chemically formed, ready to undergo still more complex changes, at the present day such matter would be instantly devoured, or absorbed, which would not have been the case before living creatures were formed.**
>
> **Charles Darwin,** letter to Sir Joseph Hooker, 1871

Panspermia

Such is the mystery surrounding the origins of life on Earth that many have suggested it was seeded from space and is perhaps widespread throughout the universe. This idea, known as panspermia, was first mentioned by the Greek philosopher Anaxagoras in the fifth century BC. It was revived by several 19th-century scientists, including Lord Kelvin, and was championed in the 20th century by the astronomer Fred Hoyle, who suggested that cometary dust might be responsible for epidemics. There is controversial evidence of micro-fossils in a meteorite from Mars; the planet may have been more amenable to life early in the solar system's history. But some argue that panspermia merely delays the problem: you still need a 'warm little pond' somewhere.

First fossils In trying to find fossil evidence of the earliest life, palaeontologists are faced with a double problem. The further back they go in the geological record, the more folded, fractured, cooked and otherwise altered are the rocks. At the same time, the creatures they are searching for are smaller and softer and less likely to leave a fossil trace.

Among the most ancient 3.8 billion-year-old rocks of Greenland are some that contain microscopic specks of carbon. They seem to contain slightly less carbon-13 than is found in inorganic carbon, and today that is taken as an indication of life. So perhaps those specks are evidence of the first life on Earth. Tiny structures in metamorphosed rocks 3.5 billion years old from Western Australia are arguably the remains of blue-green algae or cyanobacteria, while larger, layered structures resemble the stromatolites in later deposits that are known to be built up by colonies of cyanobacteria. Possible micro-fossils from 3.4 billion-year-old Australian rocks nearby come from what was then a shallow warm sea; they appear to have got their chemical energy from sulphur in the iron pyrites sand.

the condensed idea
Turning chemicals into life

38 Evolution

Once a chronological sequence of strata is established, it becomes clear from the fossils within the different layers that life has changed over time. There is one idea that makes sense of that, explaining the rich diversity of all plants and animals on Earth: evolution. Charles Darwin's theory on the origin of species through natural selection has transformed both biology and palaeontology.

Medieval Europe inherited several fallacies from Aristotle, among them that the Sun orbits the Earth and that creatures take on fixed forms reflecting a divine cosmic order. His astronomy was overturned before his biology. By the 18th century, exploration had revealed a truly vast array of life forms or species. Fossil collectors added another dimension, revealing species many of which are now extinct. Naturalists could not help noticing similarities between some species and speculating that they might be related.

Inheritance At that stage, no one knew about genes or DNA, and so the mechanism of inheritance and the means of diversification were a mystery. One of the first to suggest a theory of evolution was the French naturalist Jean-Baptiste Lamarck. In a lecture in 1800, he proposed two principles: one of increasing complexity; the other of adaptation to the environment. He speculated that characteristics acquired during an animal's lifetime might be passed on to their descendants. In that way, a muscular blacksmith was more likely to have a strong son. Similarly, traits that were not used would die out, making moles that live underground blind and birds toothless.

timeline

c.340 BC	1686	1735	1751	1798
Aristotle suggests creatures take on fixed forms that reflect divine cosmic order	John Ray introduces the concept of species defined by observable features	Carolus Linnaeus introduces binomial classification, still used for genus and species	Maupertius suggests natural modifications accumulate to make a new species	Thomas Malthus publishes his essay on population

CHARLES DARWIN 1809–82

Charles Darwin was the son of a prosperous Shropshire doctor. He neglected his medical studies in Edinburgh and found his geology lectures boring, but he developed an interest in natural history, which he went on to study at Cambridge. His wife Emma came from the wealthy Wedgwood family, which meant that Charles never had to work for a living and was able to devote himself to natural history. He could afford to pay for his passage on HMS *Beagle*, which took him on a five-year trip around the coast of South America, studying the wildlife and collecting specimens. It is probable that the trip gave him the inspiration for his theory of evolution by natural selection, but he didn't publish his famous book *On the Origin of Species* until 23 years later – perhaps because he was afraid of the reaction it might provoke in religious circles.

Survival of the fittest In 1798, a pamphlet was published anonymously entitled 'An Essay on the Principle of Population'. It turned out to be by the Reverend Thomas Malthus, who suggested that population growth would lead to a struggle for existence in which the best adapted would survive and the unfit would perish. That essay was to influence the thinking of two key men in the story of evolution: Alfred Russel Wallace and Charles Darwin.

The two men came from very different backgrounds and travelled in different ways. Darwin took the equivalent of a five-year world cruise on HMS *Beagle*. Wallace struggled to pay his way by selling specimens collected in the malarial swamps of Southeast Asia. But both were struck by the same thing: just how well-adapted plants and animals are to their specific environments. Both realized that only the fittest – those best adapted to their environments – would survive to reproduce. The concept of natural selection was born.

1800	1858	1859	1889	1953
Lamarck proposes his transmutation theory for the inheritance of acquired characteristics	Alfred Russel Wallace's theory and Darwin's are presented to the Linnean Society	Charles Darwin publishes *On the Origin of Species*	Hugo De Vries proposes the concept of genes	Crick and Watson discover the structure of DNA that carries the genetic code

Precedent On 1 July 1858, papers by both men were read at the Linnean Society in London. Wallace was still in the Malay Archipelago and Darwin had just lost his infant son to scarlet fever, so neither attended; their papers were read by the secretary. A year later, Darwin published his famous book on the origin of species, and as a result it was he who took the credit – and the flak from religious opponents – for the theory of evolution.

Evolution has proved to be more controversial than even Darwin expected. It led to a heated exchange in Oxford in 1860 between Thomas Huxley, representing Darwin, and Bishop Wilberforce, speaking for the church against evolution. Issues raised by Darwin's own cousin, Francis Galton, about the fitness of those with inherited diseases or mental impairment, led eventually to the eugenics movement and compulsory sterilization. Even today, and even in the relatively well-educated USA, many religious fundamentalists still believe that the principal types of animal and especially humans were created fully formed by God.

Evolution today The concept of evolution is still beset with misunderstandings. A widespread fallacy is the idea that humans are descended from chimps or gorillas. We are not. But we may have shared a common ancestor 6 or 8 million years ago. The fossil record is woefully

---

ALFRED RUSSEL WALLACE 1823–1913

Wallace came from a very different background to that of Darwin. His father had to move away from London to save money, and Alfred became a land surveyor in Mid Wales. He was an ardent socialist, always concerned for the plight of the poor. He had to fund his research trips by collecting specimens to sell to museums; he once lost everything from one expedition when his ship caught fire. But he persisted and, while collecting butterflies in the Malay Archipelago, devised a theory very similar to that of Darwin. He respected Darwin and wrote to him outlining the theory to ask for Darwin's reaction. It might have been that letter that prompted Darwin to rush ahead with publication of *On the Origin of Species* in 1859.

incomplete and, while it is easy to spot similarities between present animals and their extinct fossil relatives, it is a big mistake to claim a direct line of descent. The term 'missing link' is widely misused in the popular press. As more and more fossils of our distant relatives or hominins are discovered, the clearer it becomes that the tree of human evolution is more like a bush with many branches. Most branches lead to extinction, and it is all but impossible to tell which fossils lie on the path of our direct ancestry.

In human evolution and elsewhere, it is clear that nature is highly inventive and that the differences that mark out ultimate success from extinction are hard to discern.

> **The universe we observe has precisely the properties we should expect if there is, at bottom, no design, no purpose, no evil, no good, nothing but blind, pitiless indifference.**
>
> **Charles Darwin**

Convergence Critics of evolution point to complex structures such as the human eye and ask: how could that ever have come about by chance? That is a challenge for evolution, but not a challenge to evolution. The clearest proof of that is the fact that different sorts of eye have been invented independently in the course of evolution five or six times, from squids and scallops to shrimps and humans. But it is also a warning not always to assume an evolutionary relationship from a similar structure. There are many examples of convergent evolution in which a similar problem has resulted in a similar solution in unrelated species – for example, the hydrodynamic shape of a shark and a dolphin.

So often, the key to survival has been to adapt and change. But it is not always the case. Some designs endure simply by finding their niche and lying low. A classic example is a small brachiopod shellfish called Lingula, which lives quite successfully in parts of the Pacific today; an almost identical fossil is found in Cambrian rocks – 500 million years old.

the condensed idea
Survival of the fittest

39 Garden of Ediacara

In Charles Darwin's day, no one believed that there were any fossils older than the Cambrian period. Now we think differently. In the Ediacara hills of South Australia, there is a rich array of fossils dating back around 600 million years. They reveal very different times and very different creatures from any we are familiar with today.

We now know of clear fossilized remains dating back about 2.5 billion years, but they are just filamentous algae and cyanobacteria: the sort of things we would describe as pond scum today. There is little else prior to the Cryogenian glaciation, apart from one controversial set of possible worm burrows found in India and dating back 1.1 billion years.

Too old to be true? Charles Darwin and his contemporaries knew of no fossils older than the Cambrian period, which, it is now agreed, began about 542 million years ago. That was true until one day in 1957, when a Leicestershire schoolboy, Roger Mason, was rock climbing in Charnwood Forest. He came across the impression of what looked rather like a fern frond on a rock. It was in Precambrian rocks, where nobody had expected to find fossils before. But he showed it to a Leicester University geologist, who recognized that it was indeed a fossil and named it 'Charnia masoni'.

In fact, disc-shaped fossils had been found in Precambrian rocks in Newfoundland in 1868 by the geological surveyor Alexander Murray. He used them as convenient markers of a particular rock layer but, as they were below the Cambrian, never dared to suggest that they were fossils.

timeline the Precambrian

3500 Ma	2500 Ma	1100 Ma	1000 Ma	635 Ma
First evidence of fossil bacteria and possibly algae from Western Australia	First clear evidence of filamentous algae	Possible worm burrows from India	Freshwater terrestrial micro-fossils in the Torridonian of north-west Scotland	End of the Cryogenian glaciation

Stromatolites

The oldest large fossil structures are layered domes up to a metre (3¼ ft) across. These are known as stromatolites – colonies of cyanobacteria or blue-green algae. The oldest clearly identified is 2.7 billion years old, from Western Australia. It is possible that the thousands of fine layers each represent a daily cycle of growth. They disappear towards the end of the Ediacaran, possibly because too many things had been grazing off them. There are still stromatolites living today, notably in the salty, warm, shallow waters of Shark Bay on the coast of Western Australia, not far from their 2.7 billion-year-old ancestors.

The Ediacara In 1946, a young geologist by the name of Reg Sprigg was sent by the South Australian government to see if disused mines in the Ediacara hills of the Flinders Ranges could be profitably reopened. While eating his lunch he started to notice fossils resembling jellyfish, which he thought to be early Cambrian or even Precambrian. But his discovery aroused little interest: the paper he wrote was rejected by the journal *Nature*. Only later was the Precambrian age and true significance of the find recognized, and the first geological period to be created for over 100 years was named the Ediacaran. It overlaps with the Vendian, named after Precambrian fossil sites in northern Russia; other Ediacaran fossils have now been found in Namibia, Newfoundland and elsewhere.

Strange creatures at dawn Some of the best examples are to be seen on a sheep ranch in the Ediacara, about 200 kilometres (124 miles) north of

... if my theory be true, it is indisputable that before the lowest Cambrian strata was deposited, long periods elapsed ... and that during these vast periods the world swarmed with living creatures.

Charles Darwin,
On the Origin of Species, **1859**

630 Ma	610 Ma	590–565 Ma	*c.*560 Ma	542 Ma
Earliest Ediacaran fossilized embryos	First large Ediacaran fossils	Doushantuo formation in China, with well-preserved fossil embryos	Age of *Charnia masoni* from Leicestershire and Ediacaran fossils from Newfoundland	End of the Ediacaran period and fauna. Start of the Cambrian period

A glimpse of the
Ediacaran seafloor
with Charnia,
Dickinsonia,
Tribrachidium and
Spriggina.

Adelaide. They show up best just after dawn, when the low-angle sunshine makes shadows from their gentle undulations and before the flies have woken up. Some are fern-like fronds up to 30 centimetres (12 in) long and similar to Charnia. Others are circular discs perhaps 5 centimetres (2 in) across. Yet others are oval and covered with parallel wavy lines. Could these be the segments of a broad, flat, wormlike creature? Some of them are up to a metre (3¼ ft) across! Another, called Spriggina, is like an elongated version of the trilobites that were to follow in the Cambrian.

So what were these strange fossils? The circular discs are probably the hold-fasts that anchored the frond-like Charnias to the seafloor. The segmented ovals, creatures named Dickinsonia, look as if they have a front and back and may have crawled slowly, grazing on cyanobacteria on the seabed and sometimes leaving a trail in the slime. But it is all too easy to look at modern animals and say that these fossils are a bit like jellyfish, soft corals or segmented worms; a likeness may not imply any ancestral relationship.

A new kingdom? Indeed, the German palaeontologist Dolf Seilacher has suggested that Ediacaran creatures represent a whole new kingdom of life, alongside plants, animals and fungi. He calls them vendobionts and suggests that they are giant single-celled organisms with developed partitions within their cytoplasm, a bit like the quilting on a mattress. He is not convinced that they had a gut; instead, he thinks they may have absorbed nutrients through their skin or had symbiotic photosynthetic bacteria inside them.

Microbial slime The conditions in which they lived are also controversial. They are found in thin layers of silt between slabs of hard quartzite, which was once sand. Sometimes there are ripple marks in the sand, suggesting waves or currents in shallow water. The fossils often leave their impressions on the underside of the slabs, which have what has been called an elephant-skin texture. That is believed to have been left by a

Fossil embryos

The rapid explosions of diverse macroscopic life forms in the Ediacaran and the Cambrian must have started from something. Palaeontologists are now turning to micro-fossils for an answer. Many Precambrian rocks are now revealing fossilized embryos, some of them no bigger than the full stops on this page. Some of the best-preserved examples come from the Doushantuo formation in China and date from about 570 million years ago – slightly before most large Ediacaran fossils. Advanced X-ray techniques reveal the individual cells within the embryo. Many are probably the embryos of sponges or corals, but some seem to show bilateral symmetry and may be the ancestors of Cambrian arthropods, worms and perhaps ourselves.

microbial mat – a slimy layer of algae on which the creatures may have fed and which formed over their bodies and helped with preservation. If they were photosynthetic algae, that also suggests an environment of shallow water.

Dinner time What is clear from all the Ediacaran fauna is that they had no hard parts. There are no shells, no protective cuticles and, crucially, no jaws. Things like Dickinsonia were very fragile: bags of fluid tens of centimetres across, yet probably less than a centimetre thick (this is clear from the way some of them have folded over before fossilization). Clearly, there were no predators around, or they would not have lasted long. So this age has been called the Garden of Ediacara, to compare it with the Garden of Eden. As one palaeontologist has put it, 'Once something had evolved with hard mouthparts, it was dial-a-pizza time as far as Dickinsonia was concerned'!

the condensed idea
Early experiments in evolution

40 Diversification

If there is one word that best describes what has happened to living organisms over the last 540 million years, it is diversification. It begins with the so-called Cambrian explosion of wonderful life in the sea and continues as plants and animals move onto the land and come to inhabit every possible niche on the planet.

Life goes underground From its start, the Cambrian period looks very different from the tranquil slime garden of the Ediacara. The remains of the seafloor are all churned up with burrows and excavations. Where once there were wide, soft, delicate Dickinsonia grazing off the microbial mats, now the worms are hiding underground. The reason can be seen from some remarkable trace fossils – fossils that record specific events, rather than just the animals that made them. In one, a set of what are clearly tracks about a centimetre (½ in) wide and made by numerous tiny feet lead across the sea bed towards a worm burrow. There are signs of vigorous digging and the worm is no longer at home. Something has eaten it for dinner.

Evolutionary arms race That something was a creature called a trilobite: an arthropod looking a bit like a large woodlouse, whose closest present-day relative is the horseshoe crab. It has evolved a hard protein carapace and its legs and mouthparts are encased in a similar tough exoskeleton, in effect giving it jaws. But it is not alone. There is a fossil trilobite with a curved section missing from its rear end. Look closely and you see it is not a recent fracture: the wound has begun to heal and is exactly the same shape as the tough mouthparts on an even bigger arthropod named Anomalocaris, or strange crab.

timeline highlights of the Palaeozoic

542 Ma	525 Ma	510 Ma	505 Ma	440 Ma
Start of the Cambrian period. Rapid diversification in the sea	Chengjiang fauna in south-west China	Burgess Shale fauna in Canada	Beginning of the Ordovician period. Fish in the sea; first arthropods on land	Ordovician period ends with an ice age

This was a world where animals ate other animals – the start of an evolutionary arms race that continued to the dinosaurs and beyond. As arthropods were inventing armour, molluscs and brachiopods discovered how to make shells to protect themselves from hungry predators. But the arms race continued. Cambrian shells have been found with neat little circular holes in them bored by some predatory creature, we know not what.

Wonderful life The flattened remains of an Anomalocaris and its compatriots are beautifully preserved in the Burgess Shale of Canada; others, in less squashed condition, can be seen in the Chengjiang formation of south-west China. They record a sudden and amazing diversification of marine life, with all sorts of fantastic creatures that, at first sight, looked like nothing on Earth today. There's Opabina, with five eyes and a long nozzle at its front, presumably for feeding. Wonderfully named Hallucigenia has a double row of spines on one side and tentacles on the other and looks

The Burgess Shale

In 1909, palaeontologist Charles Walcott was travelling in British Columbia with his family, looking for fossils in the Canadian Rockies. The story has it that his wife's horse slipped and revealed a slab of slate packed with strange fossils. Walcott traced the slab to its source in the hillside above and returned to the site many times over the next 15 years, digging out a small quarry and collecting over 65,000 exceptionally well preserved specimens. He spent the rest of his life classifying them according to present-day animals such as crustaceans. In 1966, the Cambridge palaeontologist Harry Whittington began studying the fossils and realized their astonishing diversity, which he dubbed the 'Cambrian explosion'.

440–410 Ma	410–360	360	335
Silurian period. Coral reefs and jawed fish in the sea. Plants return on land, plus spiders and centipedes	Devonian period. The age of the fishes. Lush vegetation establishing on land	Start of the Carboniferous period after an extinction event occurs	Early tetrapods at the Black Lagoon near Edinburgh. Probable start of egg-laying reptiles

The creature from the Black Lagoon

Once, 335 million years ago, there was a tropical lagoon near East Kirkton, outside Edinburgh. It was surrounded by lush vegetation of tree ferns and club mosses, but there were frequent fires, perhaps started by nearby volcanic activity and invigorated by high levels of oxygen in the atmosphere. Land creatures fleeing the fires died and became buried in the lagoon. The oxygen enabled dragonflies, scorpions and other invertebrates to grow up to a metre (3¼ ft) long. Primitive tetrapods – amphibians of sorts – crawled out of the lagoon. One of them has been given the delightful Latin name of *Eucritter melanolimnetes* – literally, the beautiful creature from the Black Lagoon. Another, with the official name Westlothiana, is better known as Lizzie and appears to be an intermediate between amphibians and reptiles.

so strange that no one was quite sure which way up it went (in fact, it probably walked on the tentacles). Marrella seems to be all lacy legs and appendages; and Anomalocaris itself had segmented flaps for swimming and a bulbous head with two barbed appendages for drawing food to its circular mouth – which was itself first misidentified as a jellyfish. It all seems like a fantastic experiment for hopeful monsters. Arguments still rage about what was related to what and which creatures gave rise to those we know today.

Anchovy fillet In the Burgess Shale, there is an insignificant-looking creature called Pikaia. In the older rocks of south-west China, there is something rather similar by the name of Yunnanozoon. Both look like little more than animated anchovy fillets. They have what seem like gill slits and the zigzag muscle blocks we notice when eating fish. And they have what may be a nerve fibre – a notochord running down the back. These are the signature features of chordates, the phylum that includes fish, reptiles and all vertebrates, including ourselves. Back in the diversity of the Cambrian, it would take greater imagination to think that these creatures would inherit the Earth.

The invasion of land Fast-forward now almost 200 million years to the early Carboniferous. By now, the anchovy fillets have evolved into bony fish, becoming the top predators in the sea. Some have four muscular fins, perhaps originally for moving around on the seafloor. Suddenly there is a new threat and a new opportunity. Perhaps to escape from predators, they find that they can use their fins to pull themselves onto the muddy shore. Plants are there ahead of them and their rich growth has put far more oxygen into the atmosphere even than we have today. Through their skin, and perhaps by taking this oxygen into their swim bladder, they find they can breathe. For a while, their descendants are amphibious, returning to the water to breed. Eventually they are able to lay eggs on land. We call these reptiles.

> **From the beginning of life on Earth there was a connection so close and intimate that, if the entire record could be obtained, a perfect chain of life from the lowest organism to the highest would be established.**
>
> **Charles Walcott, 1894**

Of course, that sequence did not happen overnight. But there is now clear evidence of intermediate stages. A quarry at East Kirkton near Edinburgh has yielded some remarkable fossils of early amphibians and even a quite lizard-like creature – possible intermediates on the evolutionary road to ourselves.

Threats and opportunities What does seem clear is that threats can give rise to opportunities for rapid diversification. The evolution of hard parts in the Cambrian led to new strategies both for predation and for defence. The evolution of legs and the ability to breathe air in the Ordovician opened up all sorts of habitats on land. Where there are new habitats to exploit and new means with which to colonize them, evolution moves in leaps and bounds.

the condensed idea
Diversification in leaps and bounds

41 Dinosaurs

The evolutionary arms race that began in the Cambrian reached its zenith in the age of the dinosaurs. For more than 160 million years, giant reptiles ruled the planet and proved that being big can be a pretty effective way to survive. Today they star in children's books and nightmares, in spectacular museum displays and big-budget movies. But not all dinosaurs were big; some were sociable, even cute.

Dinosaurs were the kings of the Mesozoic era. They first appear in the late Triassic, about 230 million years ago. They are a wide and varied order of reptiles with more than 1,000 named species. Technically, the order excludes the big marine reptiles and pterosaurs but includes one class of descendants which is not extinct: birds.

The ultimate arms race The latest images in TV dinosaur documentaries seem to depict every species as the biggest and fiercest. They certainly have plenty to choose from. The largest are the great herbivorous sauropods, and the record holder among those is the long-necked Argentinosaurus. It was nearly 40 metres (131 ft) long and weighed in at nearly 100 tons. Competing for the fierceness prize – along with the famous theropod, Tyrannosaurus rex – is the slightly larger, crocodile-jawed, sail-backed Spinosaurus, weighing in at around 8 tons.

Dinosaurs are ahead of the game for long names and impressive statistics, but what it all adds up to is that size matters. The more powerful your jaws and the bigger your stride, the more chance you have of securing dinner.

timeline highlights of the Mesozoic

250 Ma	230 Ma	200 Ma	160 Ma
Start of Mesozoic era and Triassic period. Rapid diversification of reptiles	Late Triassic. First recorded dinosaurs	Extinction event. Start of Jurassic period	Late Jurassic. Diplodocus and Stegosaurus on land; Pliosaur and Plesiosaur in the sea

MARY ANNING 1799–1847

Mary Anning's home in Lyme Regis, Dorset, meant that she was ideally placed for collecting Lower Jurassic marine reptiles from the cliffs. She found the first ichthyosaur to be correctly identified when she was 12 years old and went on to find and identify many species including plesiosaurs and flying pterosaurs. It was a dangerous occupation: going out, often in winter, after fresh landslides to look for fossils before they were washed away by the tide. In 1833, she was almost killed and lost her pet dog in a landslide. But her gender, social class and nonconformist religion meant that it was hard for her to gain recognition among the gentleman geologists of the day, and she was never admitted to the Geological Society of London.

Even for the vegetarians: the bigger you are or the more armour plating you carry, the less chance you have of becoming dinner. It was an evolutionary arms race limited only by the ability of dinosaurs' legs and muscles to support their bulk.

Keeping warm, keeping cool Supporting your own weight is not the only problem raised by growing big. All reptiles today are cold-blooded. Actually, that is a misnomer: their temperature depends on external factors. After a cold night, a snake needs to lie in the sun and warm up before it can become active. But they can also overheat. The problem with getting big is that the proportion of surface area to bulk goes down. So if you are cold, it takes longer to heat up, and if you are hot, it is difficult to dissipate that heat.

The age of the dinosaurs was marked by a significantly warmer climate than we have today, so keeping cool may have been the bigger issue. There is evidence of extensive blood vessels in the huge plates down the back of Stegosaurus, suggesting that they doubled up

> ❝If we measure success by longevity, then dinosaurs must rank as the number one success story in the history of land life.❞
>
> **Robert T. Bakker,**
> *The Dinosaur Heresies,* 1986

150 Ma	145 Ma	125 Ma	80 Ma	65 Ma
Archaeopteryx flying around in southern Germany	Start of Cretaceous period. First flowering plants	Chinese feathered dinosaurs	Late Cretaceous. Tyrannosaurus on land; Kronosaurus in the sea; pterosaurs in the air	End of Cretaceous period. Sudden extinction of all remaining dinosaurs

Maiasaura was a herbivorous dinosaur of the Cretaceous, growing to about 9 metres (30 ft) long. They lived in huge herds. The discovery of nests of hatchlings suggests they cared for their young.

as cooling fins, like elephants' ears. There is also controversial evidence from the microscopic structure of dinosaur bones that they may have been warm-blooded – that is to say endothermic: controlling their body temperature from within as mammals do. The discovery that fine, feather-like down grew on the bodies of some dinosaurs suggests that this may have evolved for insulation and provides further evidence that they may have been warm-blooded.

Feathered dinosaurs Some of the most exciting dinosaur fossils discovered in recent years come from Liaoning Province in north-east China. Many have been beautifully preserved in fine-grained volcanic ash from shallow lakes. They show small details including, in some cases, feathers. Some dinosaurs just have a downy coating for insulation, but others have large feathers with a central quill, like those of modern birds. Many are quite small. One, named Microraptor, was not much bigger than a chicken but had well-developed feathers – on all four legs. It does not look as if it could fly; the feathers were more likely to be for sexual display. Feathers may have been developed first for insulation and then for display, before becoming used first for gliding and then full flight.

Exactly when and how modern birds evolved is still the subject of controversy. The rocks of Liaoning are around 20 million years younger than those in Germany, where Archaeopteryx was found. Archaeopteryx, discovered only a year after Darwin published *On the Origin of Species*, really did seem like a 'missing link'. It had long feathers and could clearly fly, yet it had teeth, claws on the wings and a bony tail. There are still arguments over whether or not it was an ancestor of modern birds.

There are two broad categories of dinosaur: bird-hipped, which confusingly includes the big herbivorous sauropods; and lizard-hipped, which includes the theropods that gave rise to birds! Those theropods, such as Velociraptor, walked on two legs and could probably run quite fast, so their feathered friends may have taken to the air in the same way that large birds like swans and pelicans do today – with a long run-up. However, other feathered theropods like Microraptor had long claws on their wings suitable for clambering up trees. Perhaps they first took off as gliders from the treetops.

SIR RICHARD OWEN 1804–92

Richard Owen trained as an anatomist and became interested in the comparative anatomy of animal species. Careful observation of their bones convinced him of evolutionary relationships, though he always doubted that the mechanism of evolution was as simple as Darwin proposed. Owen became interested in the huge fossil reptile bones that were being dug up in England, and in a memorable address to the British Association for the Advancement of Science in 1842 he coined the term 'Dinosauria', or 'terrible lizards', to describe them. He was the driving force behind the establishment in 1881 of what is now the Natural History Museum in London.

Cute, caring dinosaurs Dinosaurs laid eggs. Fossil evidence has been found of them nesting in colonies and sitting on their eggs. The small size of eggs and evidence of down suggests that the young were small, cute and fluffy – characteristics that we associate with producing a caring response in parents. It seems likely that some dinosaurs were social animals and not only in order to raise their young. Dinosaur gangs would have been much more effective at hunting than individuals would be if hunting alone.

the condensed idea
Survival of the biggest

42 Extinction

Ninety-nine per cent of all known species are now extinct! If you include an estimate of the fossil species that have not been identified, the proportion rises to more than 99.9 per cent. They did not go gradually. The geological record reveals five major extinction events in which more than half the species on Earth were wiped out, the most famous of which is the one that ended the age of dinosaurs 65 million years ago.

You would have needed a high-speed camera to catch this shot 65 million years ago as a 7 km diameter asteroid slams into the sea off Mexico.

First clues In 1980, Nobel Prize-winning physicist Luis Alvares and his geologist son Walter put forward a hypothesis to explain the extinction event at the Cretaceous/Tertiary boundary (the K/T boundary). They suggested it had been caused by an asteroid impact. Their evidence came from a thin layer of pale clay found at the same depth in many places around the world. The layer contains high levels of the element iridium, which is rare in the Earth's crust but abundant in asteroids. Also in the layer, especially around the Caribbean, are shocked grains of quartz and tektites – small spherical beads of glass that have solidified from molten rock in the atmosphere.

Cosmic impact Eventually, the source was tracked down to the Chicxulub impact crater, just off the Yucatán Peninsula in Mexico (see box: Chicxulub). Calculations suggest that the crater was formed by an asteroid 6 or 7 kilometres (4 miles) in diameter, coming in at a low angle faster than a high-velocity

timeline major extinction episodes

65 Ma	205 Ma
End of Cretaceous period and the dinosaurs	Triassic/Jurassic boundary. Around 55% of marine genera and most large amphibians wiped out

Chicxulub

Back in the 1960s and 1970s, geologists exploring for oil reserves had spotted what they thought might be a giant impact crater off the Yucatán Peninsula in Mexico, but the oil companies would not release detailed data and the suggestion went largely unnoticed. In the 1980s, the Alvares hypothesis set geologists on a new search which again centred on the Caribbean, where the K/T boundary layer was thickest and included jumbled deposits from an enormous tsunami wave. Seismic surveys at sea, radar on the space shuttle and samples from boreholes all homed in on the circular structure just off the Yucatán Peninsula near the town of Chicxulub. It is the remains of an impact crater 180 kilometres (112 miles) across and dating from 65 million years ago.

bullet. To the unfortunate dinosaurs it would have seemed that the sky had been split open with fire. Within a second the asteroid punched a hole nearly 30 kilometres (19 miles) deep in the Earth. It melted tens of thousands of cubic kilometres of rock which would have remained as a lava lake in the collapsed crater, perhaps for hundreds of thousands of years. The so-called low energy ejecta were thrown out with enough force to cover areas thousands of kilometres away with a thick blanket of debris. That was followed by a tsunami hundreds of metres high. The high-energy ejecta, mostly vaporized rock, would have punched a hole in the atmosphere reaching almost to the Moon before falling back to envelop the Earth, destroy the ozone layer and trigger wildfires worldwide.

As if that was not enough, the asteroid hit the sea above a thick layer of limestone and anhydrite. Vaporized anhydrite would have produced a global cloud of sulphate aerosol, shutting out sunlight and preventing plant growth for several years before raining out as sulphuric acid. Meanwhile, vaporized limestone injected carbon dioxide into the atmosphere, which warmed the climate for the subsequent few centuries.

251 Ma	375–360 Ma	450–440 Ma
Permian/Triassic boundary. Around 96% of marine species and 70% on land become extinct	Devonian/Carboniferous transition. A series of extinction pulses take out 70% of species	Ordovician/Silurian transition. Two events wipe out 57% of genera

The next extinction

So could it happen again? There is certainly no reason to think that we are now immune from asteroid impact or catastrophic volcanism. There is already a good system for spotting asteroids and tracking their orbits, but it will be a while before we have the technology to deflect them. There is slight evidence that extinctions tend to come roughly every 62 million years, perhaps as astronomical events stir up comets in the outer solar system. The last was 65 million years ago! But an extinction event may already be under way. At the present rate of extinctions, it has been estimated that we may have lost 50 per cent of all species on Earth by the end of the century as a result of human activity – hunting and habitat loss in the past and climate change still to come.

It is no wonder there were mass extinctions. But it may have been even worse! There may have been multiple impacts, perhaps from fragments of one original object. Smaller craters of the same age have been identified in the North Sea and the Ukraine, with another large but more controversial one off the west coast of India.

Volcanic outburst As deadly as an impact that size would seem to have been, there are other candidates for the cause of the extinction. One of the more convincing theories is a series of vast volcanic eruptions. Sixty-five million years ago, the Indian subcontinent was drifting over a mantle plume where the volcanic island of Réunion now lies. A rising pulse of molten magma split the subcontinent, sending part of it north to crash into Asia and leaving a section under the ocean around the Comoro Islands. The Indian half is marked by some of the biggest deposits of flood basalt on Earth, which now make up the Deccan Traps, more than 2 kilometres (1¼ miles) thick and covering an area of 500,000 square kilometres (193,051 square miles). The volcanic dust and sulphate aerosol accompanying such eruptions would have caused a significant drop in global temperatures as it reflected sunlight. That would have been followed by a rise in temperature due to the carbon dioxide emitted. The overall result would be an unstable, oscillating climate.

A question of timing There are arguments in favour of the impact theory and the volcanic theory, alongside several other theories involving climate change or falling sea level, any of which could have been bad news for the 50 per cent of genera and 75 per cent of species of plants and animals which became extinct. Most hotly debated is the precise timing. It seems clear that many species were already in decline before the impacts took place, and it is possible that the impacts occurred up to 300,000 years before the greatest rate of extinction, though such relatively small time intervals are hard to measure. Volcanic episodes had begun two million years before the K/T boundary, which could have started the decline. The consensus may be that all the theories are correct. It may take a long-term stress and a short-term shock to cause an extinction event in a biological system.

The biggest extinction of all Whatever the cause, extinction events have happened several times and the K/T event is not the biggest. That dubious honour goes to what happened at the end of the Permian period 251.4 million years ago. It has been termed 'the great dying': a time when 96 per cent of all marine species and 70 per cent of terrestrial vertebrates disappeared from the planet. No impact event has been firmly identified from this time, but there is no ocean crust as old as that, so if it had happened at sea the record would have gone. However, there was another big flood basalt episode at the time, in Siberia, the biggest known, covering 2,000,000 square kilometres (772,204 square miles) in lava.

Altogether, there have been five major extinction events over the last 500 million years, in which at least half the species on Earth were wiped out, plus at least 16 smaller events. All of them may be down to that deadly combination of a long-term stress and sudden shock.

> **If a large extraterrestrial object – the ultimate random bolt from the blue – had not triggered the extinction of dinosaurs 65 million years ago, mammals would still be small creatures, confined to the nooks and crannies of a dinosaur's world.**
>
> **Stephen Jay Gould**

the condensed idea
Extinction: all change!

43 Adaptation

The last 65 million years has been the age of mammals. Initially small, furry and warm-blooded, they were able to adapt and diversify as soon as the stage was cleared of dinosaurs. But, like dinosaurs, mammals too found success by growing big, only to suffer in their turn as the climate changed, and apes began using tools to adapt their environment.

A simple definition of a female mammal might be something with a mammary gland, that is, able to produce milk to feed its young. That's an effective definition for the present day, but milk glands don't leave good fossils, so the first mammals are defined in terms of jaws and ears! The lower jaw of all mammals is a single bone; but in all other jawed vertebrates, there are three main lower jaw bones. In mammals, the other two bones are found in the middle ear, performing an entirely different function.

Mammal-like reptiles Long before there were true mammals, there were mammal-like reptiles, or therapsids. These must have competed with the ancestors of dinosaurs and nearly won. By the late Permian, some were the size of a rhinoceros and were the dominant predators of the time. They suffered a major setback in the extinction at the end of the Permian, in which 70 per cent of all land vertebrate species vanished. It took 30 million years in the Triassic for vertebrates to re-establish in every niche and this time dinosaurs came out on top in what is sometimes known as the Triassic takeover.

timeline the age of mammals

270 Ma	248 Ma	125 Ma	85 Ma	65 Ma
First mammal-like reptiles	Permian extinction and the Triassic takeover	First distinct monotreme and marsupial mammals	Probable first true placental mammal	Cretaceous extinction event and end of dinosaurs

Even in the Triassic, mammal-like reptiles still proved adaptable. They had a bony secondary palate which probably made chewing and hence digestion more efficient and enabled them to breathe and eat at the same time. One group, the cynodonts, may have developed hair and were possibly warm-blooded and able to lactate. Some species were burrowing; up to 20 individuals have been found in one burrow system, trapped by a flash flood, suggesting that they were social animals.

First mammals Mammals themselves probably arose from cynodonts. At first they were small, nocturnal and insectivorous – like shrews. That may have helped them evade the notice of hungry dinosaurs, as well as favouring the evolution of warm blood, insulating

Megafauna

Mammals have never been caught up in an evolutionary arms race to compare with the dinosaurs, but, as the climate cooled and mammals diversified, growing large became a useful strategy. Almost every family produced giants. Among marsupials there were giant kangaroos and wombats. There were mammoths and woolly rhinos; giant short-faced bears and giant elk; giant beaver and sabretooth cats. They all tended to have long lives and few predators, but a slow breeding rate. In the last 50,000 years, most of them have become extinct, and it is tempting to speculate that human hunters were largely responsible. Certainly, most of the remaining megafauna – elephant, rhino, whales, gorillas, tigers and so on – are still under threat due to hunting, poaching or habitat destruction.

50 Ma	7 Ma	3.5 Ma	1.8 Ma	100 ka
Rapid diversification establishes main modern mammal families	Last common ancestor of humans and chimpanzees	Cooling climate stimulates human evolution	*Homo erectus* in Africa	*Homo sapiens* leaves Africa

Homo sapiens

One mammal species has done more to transform the Earth than any other: *Homo sapiens* – ourselves. Fully modern humans have been around for little more than 100,000 years, when they spread out from Africa and colonized the world. Before them came *Homo erectus* – also large-brained, bipedal and a tool user. An archaic form of modern human left Africa perhaps 1 million years ago and, in northern Europe, gave rise to the Neanderthals. Tracks show that our probable ancestor *Australopithecus afarensis* was walking upright 3.6 million years ago, and even before that there are several candidates in Africa for human ancestry.

> **In all works on Natural History, we constantly find details of the marvellous adaptation of animals to their food, their habits, and the localities in which they are found.**
>
> **Alfred Russel Wallace**

fur and a good sense of smell. The development of a sophisticated sense of smell required an enlarged brain, and that may have been one of the driving forces leading to brainy mammals. That most of them were less than 50 millimetres (2 in) long also means that their fossil remains are rare throughout the Mesozoic era.

By 125 million years ago, three main groups that are still present today had already diverged. Monotremes such as the platypus are the most primitive, producing milk but just as a sweaty secretion from a patch of skin without a nipple. Marsupials give birth to tiny young which they keep in a pouch while they suckle; whereas placental mammals such as ourselves give birth to young which have been nurtured inside the mother's body until they are much more developed.

Adaptation It seems likely that most of the major orders of mammal were already present in the Cretaceous period, but the modern families only emerged once the dinosaurs were out of the way 65 million years ago. Exactly how they are all related is still the subject of controversy and depends on whether you look at the anatomy of fossils or molecular similarities between present-day species. Either way, the range is spectacular and some of the relationships are surprising. For example, seals are related to cats and dogs, whereas whales and dolphins are closest to pigs and cattle, and the closest living relatives of elephants are dugong and manatee. As that list suggests, diversification and adaptation to every imaginable way of life has been the mammalian key. Mammals glide, climb and burrow, gnaw, graze, scavenge and kill.

Possible human ancestry: Adapis, a 50 Ma lemur; Proconsul (20 Ma); Australopithecus (2.5 Ma); *Homo habilis* (1.8 Ma); *H. Erectus* (1.6 Ma); early *H. Sapiens*; modern *H. Sapiens*.

First humans The final twist to mammalian evolution has been the emergence of our own species. It is perhaps the ultimate adaptation in that we are able to use tools and technology to adapt our environment to our needs, bringing about change much more quickly than biological evolution. Human ancestors are among the rarest of fossils, but they are so highly sought after that many candidates have now been discovered. The progressive evolution of some traits such as walking on two legs and developing a big brain is now clear. Less obvious, once you go back more than a couple of million years, is which, if any, of the many hominin species is our true ancestor. The tree of life is in fact a many-branched bush.

the condensed idea
Mammals adapted and changed the world

44 Fossil molecules

Fossils are not simply dead creatures that have been turned into stone. Sensitive new analytical techniques are revealing that some of the chemicals of life can, occasionally, remain. These molecular fossils are giving new clues to evolution and its timetable. At the same time, present-day species carry a living legacy of their ancestors in their genes.

The chemicals of life are both complex and fragile. After death, they decay over hours, days and years. But in certain circumstances, some molecules – or at least their fragments – can last for thousands, perhaps millions of years, giving archaeologists and even palaeontologists a new window on life in the past.

Fossilization Most dead plants and animals get eaten. After large scavengers and microscopic bacteria and fungi have done their bit, little is left that is not already mineralized, such as shell and bone, and even that can be eroded, dissolved or ground to dust. The creatures that are buried quickly and escape early destruction can slowly be filled and replaced by other minerals in the same sort of process of diagenesis that creates sedimentary rock.

Molecular dating Sometimes, the minerals in shell or bone trap proteins and DNA and protect them. But they will still tend to decay spontaneously at a steady rate. Unlike radioactive decay, this is a chemical process and its rate depends on external factors such as temperature. So organic molecules are not as useful as isotopes for dating. Nevertheless,

timeline most recent common ancestor with humans

460 Ma	340 Ma	310 Ma	180 Ma	140 Ma	105 Ma
Sharks	Amphibians	Reptiles, dinosaurs, birds	Platypus	Marsupials	Elephants

Genetic tracks

DNA tests can solve paternity disputes, but they can also reveal more distant ancestry. Mitochondrial DNA, passed down the maternal line, and Y-chromosome DNA inherited through the male line can reveal gender differences in ancient migration. (There is a lot of Viking DNA in the Y-chromosomes of the inhabitants of coastal north-west Europe!) There are many genetic traits or polymorphisms that confirm resistance to disease, and they can reveal both the geographical spread of disease in the past and the migration of humans from disease-affected areas. The clearest evidence of both comes from exposure to malaria, which can lead to many resistant polymorphisms. Some of these appear to have been imported into Europe, perhaps by soldiers returning from the Asian campaigns of Alexander the Great.

they are used in certain circumstances; a good example is the ostrich eggshells that litter many prehistoric archaeological sites in Africa. Many proteins can come in two mirror-image forms: left- and right-handed, as it were. In life they are all left-handed, but after death they start to decay into either-handedness, a process known as racemization. So, for given temperatures, the proportion of right-handed protein is a measure of age.

Ancient genes DNA is a fragile molecule and quickly breaks down into short fragments by a process of hydrolysation. But fragments can be preserved in shell, bone, tooth or other impervious materials such as amber. As with protein decay, the speed of breakdown depends on temperature, so, for example, the remains of mammoths preserved in Siberian permafrost stand more chance of containing useful DNA than, say, bones of the same age from the Sahara Desert.

85 Ma	75 Ma	63 Ma	40 Ma	18 Ma	14 Ma	7 Ma
Dogs, horses, whales	Rodents, rabbits	Lemurs	New World monkeys	Gibbons	Orangutans	Chimps and bonobo

Modern techniques, in particular the polymerase chain reaction, can make thousands of copies from a single fragment of DNA that can then be sequenced. But the sensitivity of the process makes it highly susceptible to contamination. Easiest to isolate is mitochondrial DNA, as each cell contains many copies in small loops in the cytoplasm. Nuclear DNA is harder. Nonetheless, sufficient mammoth DNA, both mitochondrial and nuclear, has been recovered to investigate their relationship to modern elephants. Strangely, mitochondrial DNA, which is passed on down the female line only, suggests a close relationship to Asian elephants, whereas nuclear mammoth DNA is closer to that of African elephants!

Bringing back the dead If enough DNA can be found to reconstruct its complete genome, it should in theory be possible to resurrect an extinct species. But it will not be easy in practice. It took a Japanese team more than 1,000 attempts to clone seven viable mice using cells taken from a dead mouse that had been in a laboratory freezer for 16 years. With older specimens stored at higher temperatures in museums, it will be even harder. But DNA has been isolated from museum specimens of extinct creatures such as the dodo, the quagga and the thalacine or

The real Jurassic Park

In the *Jurassic Park* films, as so often in science fiction, plausible science is taken to implausible limits. The films suggest that bloodsucking insects that fed on dinosaurs might get trapped in amber, preserving dinosaur DNA in their stomachs, and from that DNA the complete genome could be reconstructed and dinosaurs cloned. In practice, it has not been possible to reproduce claims of extracting even insect DNA from fossil amber, let alone dinosaur DNA. Even if there is DNA, it would be broken down into such tiny fragments that reconstructing the chemical jigsaw would be impossible, even with sophisticated supercomputers. Perhaps fortunately, *T. rex* is extinct for ever.

marsupial wolf. The same team that cloned the frozen mouse has suggested that it may be able to recreate a living mammoth from frozen DNA within five years. Doing so would involve replacing the DNA in an elephant cell with mammoth DNA, inserting it into an elephant egg cell and implanting that into the womb of a living elephant.

Living fossils Fossil DNA is rare, but every plant and animal contains living molecular fossils of its ancestors. Anyone who doubts the evolutionary relationship of living things need only look at the similarities between the genes of different species. Functional genes that perform the essentials for life are conserved across a vast range of species. Not only do humans have genes in common with chimpanzees, we also share a few genes with fruit flies and even yeast. About 50 per cent of our genes have counterparts in bananas!

Molecular clock The genome of a species also contains long sequences of so-called junk DNA, with no apparent purpose. Mutations in these regions therefore get passed on and gradually accumulate like the ticks of a molecular clock, making it possible to estimate how long ago two species diverged. The result does not give an absolute age unless calibrated with fossil evidence, but it can suggest, for example, that humans and orangutans diverged twice as long ago as humans and chimps. To put an estimated date on that, the most recent common ancestor of chimps and humans lived about 7 million years ago. The evolutionary relationships according to the molecular clock are in broad agreement with those deduced from anatomy, but there are sufficient differences to have resulted in heated debate.

> **Almost all aspects of life are engineered at the molecular level, and without understanding molecules we can only have a very sketchy understanding of life itself.**
>
> **Francis Crick**

the condensed idea
Evolution revealed by molecules

45 Anthropocene

Since the last ice age ended 11,700 years ago, we have enjoyed a relatively stable climate and an epoch that geologists call the Holocene. This has made agriculture, cities and global trade possible. But some geologists now believe that human activity is so irrevocably altering our planet that we have entered a new geological age: they call it the Anthropocene.

The term Anthropocene was coined by ecologist Eugene Stoermer and popularized by Nobel Prize-winning chemist Paul Crutzen in the year 2000. He argued that humanity has changed the world so much that there will be an unmistakable division in the geological record, constituting a new epoch.

Entering a new age Past epochs in the geological record have typically lasted for around 10 million years, so, at little over 10,000 years, the Holocene has been very short and some have suggested it should simply be renamed the Anthropocene. That would fit reasonably well with the rise of agriculture in the Neolithic, when forests were first felled to make way for farming. But that in itself did not change the natural world greatly.

What geologists look for to mark a new epoch is what they call a 'golden spike' – some distinct marker that will show up in any sort of rock anywhere in the world. One possible marker comes about 2,000 years ago, when the Romans began large-scale lead mining and smelting, leaving a

timeline humans in the geological record

2.6 Ma	7000 BC	1st century AD
First widespread stone tools (Oldowan) in East Africa	Widespread forest clearance, first cities	Rise of lead in sediment cores due to smelting

clear trace of the metal that has reached as far as ice cores from Greenland. Another suggestion is around the year 1800, being the approximate start of the industrial age. That is reflected in a rise in mercury levels in sediment and ice cores, due to mercury being released from burning coal. It is accompanied by a rapid rise in the human population, along with the beginnings of the current rise in atmospheric carbon dioxide.

Another suggestion is that the Anthropocene should begin in 1945, at the end of the Second World War. That marks another step in population rise and urbanization. It is also a geological horizon that should be easily recognizable in sedimentary deposits millions of years in the future, because it marks the start of the nuclear age. The nuclear bombs detonated over Hiroshima and Nagasaki and the atmospheric test explosions that followed will have left their mark in the radioactive isotopes now buried worldwide in layers of mud of that age.

Era, period, epoch or age?

There is a hierarchy to the divisions of geological time. The Precambrian, which lasted nearly 4 billion years, is a super eon and contained three aeons. The fourth and most recent eon, the Phanerozoic, has lasted for the last 542 million years. It is, in turn, divided up into three eras: the Palaeozoic, Mesozoic and Cenozoic. They each contain several periods, such as Jurassic or Cretaceous, which are further subdivided into epochs, typically of around 10 million years or so. Our present Quaternary period contains the Pleistocene and Holocene epochs. The question is: does the Anthropocene represent such a major change that it should mark the start of more than an epoch – perhaps a period or even an era?

1800	**20th century**	**1945**	*c.***1970**
Atmospheric CO_2 levels start to rise	A second peak of lead in sediments, from car exhausts	First radiogenic isotopes from atomic bombs in sediment and ice cores worldwide	Plastic fragments start to be common in sediments

> **‘I was at a conference where someone said something about the Holocene. I suddenly thought this was wrong. The world has changed too much. No, we are in the Anthropocene. I just made up the word on the spur of the moment. Everyone was shocked. But it seems to have stuck.’**
>
> **Paul Crutzen**

Degrees of change The beginning of the last geological period, the Quaternary, was marked by the onset of a series of ice ages. The beginning of the present era dates to the extinction of the dinosaurs and rapid climate change. So, do the changes being brought about by the human species add up to something that will stand out as clearly in the geological record of the future as either of those?

Mass extinction Compared with the slow and relentless changes mediated by James Hutton's deep time, the last 70 years have been sensational. The period has been referred to as the 'Great Acceleration'. In that time the world population has more than doubled. Carbon dioxide emissions have multiplied sixfold. Average temperatures have begun to rise, as has sea level, while many glaciers have retreated. The biomass of algae in the oceans has reduced by 40 per cent. Some natural habitats have been reduced by 90 per cent, and the rate of species extinctions is between 100 and 1,000 times faster than the background rate, perhaps as rapid as it was at the end of the Cretaceous. Seen in a geological context, the early Anthropocene will represent one of the great extinction events of all time.

Human remains So, assuming technology doesn't take over the planet to the extent of stopping normal geological processes, what clues will our civilization have left in the rocks in 100 million years time? There will be the evidence of climate change, extinction of species and loss of biodiversity, plus the tell-tale isotopes left by our nuclear industry. But what of our monuments, cities and homes?

Fossil cities

Abandoned cities that remain exposed above ground will eventually be eroded, entering the rock cycle as particles of sediment. But the foundations below ground or cities lost beneath the waves may become buried and fossilized. In 100 million years, what will remain? Iron will rust; wood will decay or carbonize. Brick might soften and turn grey as the firing process reverses, and concrete will begin to crumble. If the remains are buried deep enough, heat and pressure will begin to change them. Plastics might revert to oil. Brick may become like a metamorphic schist and concrete may turn to marble. Eventually, all substances may melt into granite, losing all traces of the hand of man.

Most geological deposits are laid down by water and yet we live most of our lives on land. Marine sediments might contain the occasional glass bottle dropped overboard and perhaps a few shipwrecks. On land, the relentless forces of erosion will take their toll. Even bricks and concrete will be reduced eventually to sand and gravel – though it will be strange sand. In addition, almost all sand being deposited today contains a small fraction of ground-up, sand-sized particles of something new to sedimentary rocks – plastic.

But some fossil cities will remain. Cities such as New Orleans, Amsterdam, Venice and Dhaka are built at or even below sea level, in regions where river deltas are building up thick sediments, causing the ground to subside. Even without subsidence, rising sea levels will eventually engulf low-lying towns and cities, burying them under mud and preserving them for geologists of the far future.

the condensed idea
Humans make their mark

46 Future resources

There are now more than 7 billion people on Earth. It has been estimated that if all of them were to attain the lifestyle of the average American, it would take five or six planets like the Earth to support them sustainably. So how can the human species live comfortably, sustainably and within its means?

With the exception of sunshine, everything we depend upon comes from the Earth: from the food we eat and the clothes we wear to the materials with which we build our homes and the energy that drives our transport, it all comes from the ground. We are an ingenious species and no doubt new technologies will emerge to extract more from our planet and make the most of what we have. But it seems clear that we live in the best of times – peak oil and perhaps peak everything.

> **'The conservation of natural resources is the fundamental problem. Unless we solve that problem, it will avail us little to solve all others.'**
>
> **Theodore Roosevelt, 1907**

Life's essentials It has been suggested that humans are about 10,000 times more common than they would be if they were just a non-technological hunting and foraging species. The species passed that threshold when farming began in the Neolithic and since then, apart from a few setbacks due to plagues such as the Black Death, the population has continued to climb. Around 1800 it reached 1 billion, and by 1927 that had doubled. It doubled again by 1974, and in 2011 the total reached 7 billion people.

timeline when will they run out?

13 years	29 years	30 years	40 years	45 years
Indium (used in LCD screens)	Silver	Antimony	Proven oil reserves	Gold

Mining the oceans

As conventional mines become exhausted, new mining techniques will have to be developed. Vast areas of the deep ocean floor are covered with nodules rich in certain elements, such as manganese and cobalt, and there are already proposals to mine these, perhaps with long underwater suction pipes. Seawater itself contains valuable minerals, albeit in very dilute quantities; it's just a question of extracting them. Just 3 per cent of the lithium in seawater would be enough to provide every family on Earth with an electric car.

All that has only been possible through agriculture. Although many individuals are still malnourished, major famines are rare thanks mainly to improvements in crop breeding, fertilizers and pesticides. But all this comes at a cost and it can't go on forever. Thirty per cent of the land surface, including most of the areas best suited to agriculture, has already been cleared for farming, and in some of those areas intensive cropping, chemicals and irrigation are wearing the soil out. As prosperity grows, more people want a diet based on animal protein, which takes more land to produce – and more water. The provision of fresh water may soon become one of the biggest political issues in parts of the world.

Rare and precious elements New technologies, particularly in the electronics industry, have put fresh demands on the supplies of relatively scarce elements. For example, liquid crystal display screens require indium; some of the newer types of solar cell need gallium; the best magnets for wind turbines and electric car motors use neodymium. Tantalum is used in mobile phones and terbium in the fluorescent coating of light bulbs.

59 years	**61 years**	**67 years**	**116 years**	**120 years**
Uranium	Copper	Natural gas (excluding hydrates)	Tantalum (used in electronics such as mobile phones)	Coal

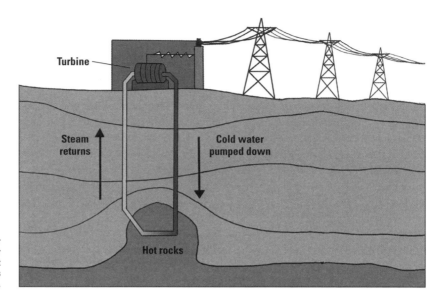

Turbine

Steam returns

Cold water pumped down

Hot rocks

A geothermal power station injects cold water down a borehole where it heats to steam and rises to drive a turbine.

Many of these so-called rare earth elements are indeed rare and hard to extract. Many of them are mined in China, and as China's own industry develops, smaller amounts are available for export. The British Geological Survey has published a risk list in which elements are scored depending on their scarcity, their geographical range and the political stability of the countries where they are mined. At the top of the risk list are antimony, platinum, mercury, tungsten and a number of rare earth elements.

Changing demand If certain new technologies become widespread, demand will suddenly outstrip supply of particular elements. For example, if electric cars become popular, there will be high demand for lithium to make their batteries and neodymium to make the magnets in their motors. One Toyota Prius currently contains nearly a kilogram of neodymium. If cars powered by fuel cells are mass produced, there will be a sudden run on platinum. If all the world's vehicles used fuel cells, platinum reserves would last no more than 15 years. As it is, a lot of platinum is used in catalytic converters for car exhausts. Much is lost in roadside dust, which can contain 1.5 parts per million of platinum, making it almost worth mining.

Fossil fuel At the present rate of consumption, proven oil reserves will run out in about 40 years' time. If we have already reached peak oil, consumption may begin to decline, making the supply last longer. New reserves will be found, but they are likely to be increasingly difficult and expensive to extract. As the price of oil rises, so more reserves become economical. Even so, we will be looking at a world without oil next century.

Natural gas reserves, too, will be used up this century. But if safe, economic ways can be found to recover methane hydrates from the seafloor, they could provide gas for a further 100 years. At present rates of extraction, coal reserves will be good for 120 years.

Mining space

Getting into space is very expensive in terms of energy, but at least on the way back you have gravity on your side! Catch the right sort of small asteroid and tow it back to Earth and, if you could get it back to the ground without it vaporizing on impact, you might have all the platinum and heavy metals you would need for a century. By the time we can do that, it might be more use to keep those elements in space to construct new spaceships and habitats. Energy for the future might come from helium-3 mined on the Moon, or even clean hydrogen scooped up from the solar wind.

Nuclear fuel The present generation of nuclear power stations relies on uranium as the main fuel, and there is only enough in existing mines to last another 60 years. Once again, further deposits may be found, but already some analysts are suggesting that power stations should be developed that can run on more abundant thorium, which, conveniently, also produces less radioactive waste. Ultimately, nuclear power stations may need to harness the same processes that power the Sun: nuclear fusion. Suitable fuel elements can be extracted from seawater, though they are only present in trace quantities. Apollo 17 astronaut Harrison Schmitt has even suggested nuclear fusion fuelled with helium-3, strip mined from the dusty surface of the Moon!

the condensed idea
Our consumption is unsustainable

47 Future climate

Climate change is the hot topic of our times. Some believe the worst predictions of the climate models developed by scientists; a few remain sceptical. But a geological perspective on both past and future climate suggests that change is inevitable: it is just a question of how much and how quickly – and whether there is anything we can do about it.

A quick glance at the geological record (see chapter 31) reveals that for long periods in the past, the Earth's climate has been in very different stable states to our own. There have been ice ages with average temperatures 7 or 8 degrees Celsius below what we now experience, and there have been prolonged warm spells, such as during the Mesozoic, with average temperatures 10 or 15 degrees higher. The differences in the past have been due to a combination of factors, including changes to the Sun, changes in the Earth's orbit, and rising or falling levels of greenhouse gases in the atmosphere. But there has never before been a time when, in the space of a century, a large fraction of the planet's fossil fuel has gone up in smoke.

The evidence mounts The evidence that carbon dioxide levels in the atmosphere are rising is incontrovertible. The physics of the greenhouse effect is well established. What is harder to prove is the exact degree to which global average temperatures are likely to change as a result. The story so far can be seen in the infamous 'hockey stick' graph, so-called due to the sudden upturn in average temperatures over the last 50 years. The graph goes back 1,000 years, and before about 1850 it has to rely on techniques other than thermometers, so some dispute its accuracy. But almost all climate scientists agree that the world is warming.

timeline future climates

Now	2012	2100	2200
Temperatures close to normal for past 10,000 years	Temperature beginning to rise (+0.8°C)	CO_2 double preindustrial levels. Temperature +3°C. Sea level +0.5 m	Western Antarctica begins to collapse. Sea level +3 m

Models and projections

Projecting that warming process into the future is complex. It takes a supercomputer to forecast the weather just a few days in advance, so predictions for the rest of the century and beyond are fraught with difficulty. Increasingly detailed models of the Earth's climate system have been run in computers with multiple small variations. The precise outcome varies, but in almost every case warming continues, climbing to a rise in temperature somewhere between 2 and 4 degrees Celsius by the end of the century.

> **We can't negotiate the facts ... It is wrong for this generation to destroy the habitability of our planet and ruin the prospects of every future generation.**
>
> **Al Gore, December 2008**

Thawing the poles

One of the uncertainties is the extent of positive feedback: that is, the extent to which a small rise in temperature might trigger something that leads to an even greater rise in temperature. Such processes can include disrupting ocean circulation and releasing methane from gas hydrates or from carbon in the Arctic tundra. Methane is 25 times more potent than carbon dioxide as a greenhouse gas. Recent estimates suggest that Arctic permafrost is thawing faster than expected and could release sufficient methane to double the impact on our climate of all global deforestation.

Masking reality

A few scientists and their many supporters have suggested that the situation is not as bad as the Intergovernmental Panel on Climate Change (IPCC) claims. They point out that, in the geological record, big increases in carbon dioxide seem to lag behind rises in global temperature. But that could be due to positive feedback: a rise in temperature could cause a rise in atmospheric carbon. The initial temperature rise in the geological past may indeed have been due to other factors, such as the Sun – no one was burning fossil fuel back then – but that is not to say that it doesn't work the other way round. In fact, global dimming due to pollution and low solar activity in the current sunspot cycle may both be masking the true extent of warming.

2500	5000	10,000	55,000
Temperature peaks at +8°C	Last remnant of Greenland ice cap melts. Sea level +12 m	Eastern Antarctica melting. Sea level +70 m	Potential ice age avoided due to enhanced greenhouse effect

Geo-engineering sunshine

The volcanic eruption of Mount Pinatubo in 1992 gave scientists an idea. For the next two years, global temperatures were about 0.5 degrees Celsius cooler, due to the sulphate aerosol the volcano injected into the stratosphere reflecting sunlight back into space. Reproduce that with several artificial volcanoes constantly injecting reflective particles from hosepipes suspended by stratospheric balloons and you might keep global temperatures stable. That is the theory, but the practice raises several unknowns and a host of ethical dilemmas.

Extreme events The climate models suggest that global warming is not going to be even. The fringes of the Arctic and Antarctic have warmed more in recent years than anywhere else. Elsewhere, climate change in many cases seems to be making existing problems worse. In tropical regions, rains are getting more erratic and deserts are becoming drier; while in temperate regions, there seem to be more of both droughts and storms. The consequences for world food production could be severe.

The rising tide Climate change will also affect the oceans. As the surface warms, the water is less able to sink down to complete the conveyor belt of ocean circulation. It is this circulation that keeps maritime climates mild, as in the UK. It also transports the nutrients essential for marine life and fisheries. Increased carbon dioxide will dissolve in the upper layers of the oceans, making them more acidic, dissolving coral and shell and harming marine life further. And finally, as polar ice caps melt, and cool water warms and expands, sea level will start to rise. Even a rise of a few centimetres will be bad news for low-lying regions facing a storm surge, and a rise of a metre would put some island states, coastal cities and even whole countries (Bangladesh, for example) at extreme risk of devastating floods. If all the polar ice melts, we'll be in for a rise in sea level of 70 metres (230 ft)!

Geo-engineering the atmosphere

So far there is little indication that nations or individuals are prepared to make the sacrifices necessary for a dramatic reduction in carbon emissions. Perhaps technology can compensate? Initial experiments with fertilizing the ocean so that plankton would draw down carbon dioxide started well, but the carbon dioxide was soon released back into the atmosphere. Extracting carbon dioxide from power station chimneys and pumping it back down disused oil wells is technologically feasible but economically unattractive. Other ideas include a concrete forest of tubes in cities that chemically absorb carbon dioxide, or converting limestone into lime and scattering it on the oceans – a process that absorbs twice as much CO_2 as is emitted in making the lime. All these geo-engineering techniques cost money and are only a temporary substitute for controlling emissions.

Beyond 2100 To the deep time of geology, a generation is insignificant, the term of office of a government doubly so. So what is in store for the climate in a century or even another millennium? Most of the climate models run up to 2100, but the problem won't end there. One scenario sees carbon dioxide levels climbing until 2050. That would bring a temperature rise of 2 to 4 degrees Celsius, which would persist for several centuries. But a worst-case scenario in which we continued until all the coal reserves were burnt would lead to a 6 to 10 degree temperature rise next century, which would persist for thousands of years and lead to complete polar melting.

the condensed idea
Lessen the inevitable

48 Future evolution

Half a billion years of plant and animal life on Earth have seen the evolution of some remarkable creatures: from microscopic to monstrous, from graceful to bizarre. The ones most suited to their environment and able to adapt as it changed are the ones that have survived and evolved. Does that process still continue? Are we humans still evolving?

Evolution, it has been said, is the product of genetics plus time. But, as Charles Darwin recognized, it also requires the process of natural selection – the survival of the fittest, or, more accurately, the reproduction of the fittest. Evolution continues as long as there is genetic variation affecting reproductive success.

While external factors are stable, evolution is a relatively gentle arms race. It might be an internal competition within a species for reproduction: a dominant male perhaps preventing other males from breeding. It might be competition from predators: staying sufficiently inconspicuous, fast or large to avoid being eaten. Or it might be driven by disease: developing sufficient resistance to fight off or at least not be killed by an infection.

The next extinction Then something comes along that moves the goalposts. It may be sudden climate change, or the consequences of an asteroid impact or major volcanic eruption. For less versatile creatures, the change is too rapid for adaptations to evolve and extinction results.

timeline completely fictitious future scenario

2025	2030	2042	2048	2050
Genetic samples from 1 million species now held in gene banks	Gene therapy for some inherited diseases routinely available	Giant pandas and Bengal tigers extinct in the wild	Cloned dodos reintroduced on Mauritius	Germline gene therapy licensed for humans

> *The dinosaurs disappeared because they could not adapt to their changing environment. We shall disappear if we cannot adapt to an environment that now contains spaceships, computers – and thermonuclear weapons.*

Arthur C. Clarke

That happened 65 million years ago and wiped out the dinosaurs. It may be happening again now; but most of it seems to be due to hunting and habitat destruction by humans and, in isolated environments, predators introduced by humans.

Already, the planet has lost many of the megafauna of the Pleistocene. Mammoths, giant elk, sabretooth cats, giant moa and many others are extinct. Others, such as giant panda, tigers, elephants, rhino and several whales, are under threat. These are all large animals with slow reproduction rates and long life spans, making them slow to evolve and vulnerable to change. Perhaps some of them would have become extinct without the added burden of human pressures. As it is, several are only being kept alive through captive breeding programmes.

Saving species Advances in laboratory genetics and cloning raise the possibility of preserving a species not as living animals but as frozen cells that might one day be allowed to grow again. Perhaps even recently extinct creatures could be resurrected in this way. Already, many thousands of plant species are stored in seed banks and scientists are racing to discover and preserve the genetic diversity of the planet before it is lost for ever. But a freezer full of cells may seem poor compensation for the loss of a forest or coral reef. If the present rate of extinctions continues, our present era will stand comparison with the great extinctions at the end of the Cretaceous and Permian periods.

2056	**2112**	**2113**	**2417**	**3642**
First company to offer cosmetic genetic modification for designer babies	Haemorrhagic influenza kills 40% of population	World rat population is double human population	Ant colony found capable of advanced chemical information processing	First evidence of giant grey squirrels using stone tools

Human evolution But what of human evolution? Our species seems to have come a long way since its ancestors split off from the ancestors of chimpanzees 7 million years ago. But the genetic changes have been comparatively few. The biggest differences may have come about through social evolution and the associated development of big brains. Some of the more significant genetic changes have been things we have lost, such as thick body hair and, thanks to the development of cooking, the ability to digest raw food. We've made some gains too. In equatorial regions, we developed skin pigments to protect our hairless bodies from sunburn. (At high latitudes, we had to lose the pigment in order to make enough vitamin D.) We've gained a gene called FOXP2 which seems essential for language. And we've continued to evolve in the battle against disease. Many African populations have a gene for a blood factor known as the Duffy antigen, which protects against a common form of malaria. And the descendants of the Neolithic farmers of Europe have retained the ability to digest milk, with a gene that normally switches off after weaning.

Future human evolution It is not easy to gauge the extent to which human evolution will continue in the future. Now that we can use the quick fix of technology to adapt our immediate environment to our needs, perhaps there is less pressure to evolve better adaptations

Brave new babies

Genetic science has advanced so far that it is easy to imagine designing genes from scratch to improve on nature or perform novel functions. It may soon be possible to correct inherited diseases – and that raises the possibility of genetically modified humans. At the moment it is only being proposed as a therapy for replacing defective genes in an individual. But in principle it could modify germ cells, the cells that give rise to eggs or sperm and hence to the next generation. That might wipe out an inherited disease that has plagued a family for generations, but it could also lead to a brave new world of designer babies.

After we are gone

If the human race was wiped out by plague or warfare, what would evolve to take our place? A few million years ago, there were several moderately intelligent hominins waiting in the wings, but now we have no obvious heirs. First to fill the void might be opportunists, the animal equivalents of weeds – perhaps rats or cockroaches. But such things live off the rubbish of human civilization, which would soon vanish. Chimps or gorillas don't appear to be in a hurry to take over, so perhaps intelligence would develop among social birds such as crows. Or, just as multicellular animals took over from protozoans, colonies of ants or termites might develop into intelligent superorganisms.

to our environment. But with intercontinental travel, cosmopolitan cities and mixed marriages, new combinations of genes are constantly coming together. As long as there are children dying before they reach reproductive age and as long as some populations are having more babies than others, natural selection will still be at work.

There may be a further big change in evolution on the horizon. Already, we are genetically engineering crop plants and even animals, both to make them more useful for their conventional applications and, in some cases, to engineer them for new uses such as the production of pharmaceuticals. In some cases, this is little different in practice from precision crop breeding; but for some, making a rabbit that glows in the dark or a tomato containing genes from a fish raises serious ethical questions.

the condensed idea
Natural selection or artificial evolution?

49 Future continents

Slowly, very slowly, your atlas is going out of date. Oceans continue to open and close; land masses collide in the ongoing intercontinental waltz. Existing maps will be good for a generation or two, but in a million years the Atlantic will be 10 kilometres (6 miles) wider. In 250 million years, it may not be there at all.

This has been going on for 3 or 4 billion years already. Lost oceans of the past have sunk deep into the Earth's mantle and ancient collisions between tectonic plates are marked by mountain ranges. So what is in store for the future?

Future oceans One of the most prominent features on a map of Africa is the Great Rift Valley. It runs like a gash through the continent from Mozambique in the south, dividing to encompass Lake Victoria and all the great lakes of East Africa, then on northwards through Ethiopia and Eritrea and into the Red Sea. Still it continues, through the Dead Sea and Jordan Valley, until it reaches Lebanon. It represents a potential new ocean.

For most of its length, the Rift Valley is a classic continental rift: a split in the continent into which slabs of continental crust have collapsed, forming a series of giant steps up the sides. But as you head north through Ethiopia,

timeline a brief history of the future

Present	50 Ma	150 Ma
Atlantic opening slowly. India completing collision with Asia	Atlantic at its widest, subduction starting. Africa moving towards Europe creating Mediterranean Mountains. Los Angeles heading north past Vancouver	Atlantic narrowing. Antarctica colliding with Australia and Borneo. Africa-Europe collision has pushed the British Isles into the Arctic

> **With such wisdom has nature ordered things in the economy of this world, that the destruction of one continent is not brought about without the renovation of the earth in the production of another.**

James Hutton, *Theory of the Earth,* 1795

towards the region known as the Afar Depression, the character changes and volcanic activity along the centre of the rift becomes more frequent. The volcanoes themselves are not tall, conical mountains but fissures that spew out runny basalt. This is much more like a mid-ocean ridge, except that it is still on land.

Still further north, in the Danakil Depression on the border with Eritrea, the ground surface is 100 metres (328 ft) below sea level. This has been described as the cruellest place on Earth: a desert with searing daytime temperatures, prickly scrub, jagged volcanic rocks and armed tribesmen. In the middle is the Erte Ale volcano, with a lake of magma that has been kept molten for a century. It lies at a three-way junction between the Rift Valley, the Red Sea and the Gulf of Aden and sits on top of a side branch of the mantle plume rising beneath Africa. That plume is trying to tear the continent apart and create a new ocean. The rift is widening and sinking. Perhaps one day the Horn of Africa will find itself separated from the rest of the continent by a widening ocean.

Pangaea Ultima. How the globe may look in 250 million years' time. The Atlantic has closed, Africa has moved north and the Indian Ocean is landlocked.

250 Ma

Pangaea Ultima. The Americas are wrapped around Africa with South Africa where the Caribbean used to be. Australia/Antarctica approaching Chile. Remnants of Indian Ocean now a landlocked sea

2,000 Ma

Earth's outer core freezes solid. Magnetic field ceases. Plate tectonics slow and eventually stop

Pangaea Ultima

There are two very different scenarios for developing a supercontinent in the future. They depend on how long it is before the Atlantic begins to take a dive. In both cases, the Atlantic continues to widen for perhaps another 50 million years, extending by a further 500 kilometres (311 miles). If the western edge of the Atlantic then starts to dive down beneath the Americas, the Atlantic will begin to close again over the next 200 million years, creating a supercontinent that has been christened Pangaea Ultima by Columbia University geologist Chris Scotese.

Vanishing oceans There's no such thing as an ancient ocean. By the time ocean lithosphere is 180 or so million years old, it has become so cold and dense that it has no choice but to sink back down into the mantle. The Mediterranean is all that remains of the mighty Tethys Ocean of the Jurassic, and even that will be gone in another 50 million years. The youngest of the great oceans, the Atlantic, will not last forever. Eventually, one or other of its margins, probably the western margin where it meets the Caribbean and the Americas, will form a deep trough and take a dive beneath the continent. Then the Atlantic will start to close again.

Future mountains The remains of the cities of the Eastern Seaboard – Boston, Rio, New York and the rest – will find themselves lifted in a volcanic mountain range like the Andes. Meanwhile, the collision of India with Tibet will slow and stop, but the northward progress of Africa towards Europe will continue until there is a long ridge of high mountains like the Himalayas where the Mediterranean once was.

The Wilson cycle Tuzo Wilson, that architect of plate tectonics, realized that the continental distribution we see today resulted from the break-up of a single supercontinent called Pangaea. He also recognized that that was not the start of the process and that generations of

Amasia

If the Atlantic does not close, it will be the Pacific that takes up the slack, giving rise to a very different continental jigsaw and effectively turning the previous supercontinent of Pangaea inside out. In this scenario, the Americas swing round into East Asia to form a supercontinent named Amasia (a name coined by Harvard geologist Paul Hoffman, combining America and Asia). Either way, the waltz goes on.

supercontinents had formed and broken up again in previous cycles, each lasting half a billion years or more. This so-called Wilson cycle is not yet at an end: the continents will come together again – the only question is whether this will be back to front or inside out (see boxes).

When the Earth freezes Plate tectonics and its associated volcanoes is how our planet loses internal heat. Much of that heat is still being generated by radioactive decay and by the inner core slowly solidifying. Four billion years ago there must have been so much heat that the mantle would have been constantly churning with volcanic eruptions worldwide, leaving little chance for stable plates to form. The evidence suggests that plate tectonics got started perhaps 3 billion years ago. In the future, it will inevitably slow down as the planet cools. Perhaps another 2 billion years and the Earth's core will have frozen solid, meaning that we will lose our magnetic field and with it, perhaps, continental drift.

the condensed idea
Continents continue drifting

50 The end of the Earth

The universe may seem a vast, unfriendly place for a small planet populated by delicate, carbon-based life forms. We are fortunate to have had hundreds of millions of years of comparative stability in which intelligent life could develop, study its home planet and marvel at it. But it can't go on forever. One day, our world will end.

Every year, there are natural disasters – earthquakes, volcanic eruptions, tsunamis, hurricanes and so on. Their local or regional impact may be tragic, but they do not threaten our species or our planet. Even the violent or prolonged eruptions of supervolcanoes or flood basalts associated with mass extinctions in the past would see at least a few survivors. For signs of the end of the world, we need to look beyond our planet.

Apocalyptic prophecies If you are reading this after December 21, 2012, then predictions that the world would end on that date were obviously greatly exaggerated. They were based on the ancient Mayan calendar, which, like all calendars, describes cycles. The Mayan cycle or 'Long Count' is longer than most, completing at the winter solstice in 2012 after 5,125 years. Although the Mayan texts make no mention of the world ending then, it has given rise to apocalyptic prophecies and a spectacular Hollywood film (*2012*).

timeline the possible future of Planet Earth

500 Ma	800 Ma	900 Ma	1,000 Ma
Terraforming of Mars begins	Colonies established on moons of Jupiter and Saturn	First interstellar ark spaceship sets out for new worlds	Increased solar radiation starts to boil the oceans

Deflecting Armageddon

If an asteroid was spotted heading towards the Earth, it might still be possible to avoid it. Blasting it to pieces with a missile is not the answer. That would just multiply the problem – and lead to some interesting insurance claims. The best bet is to spot the asteroid several orbits in advance: then only a gentle nudge would be needed. Spraying one side with reflective paint might enable sunshine to do the rest. A rocket engine, perhaps even a gentle electric propulsion system, if landed on the asteroid, could produce enough thrust to move it into a new orbit. If all else failed, detonating a nuclear explosion, not on the asteroid but nearby, might deflect it without breaking it up.

The prophecies talk of earthquakes and tsunamis, solar storms and planetary alignments. The planets are not due to align until 2040, and when they do, the tidal effect is only 64 millionths of that of the Moon. There is also talk of a collision with a mysterious planet called Nibiru. That was supposedly discovered by the Sumerians around 2500 BC and is said to have a highly elliptical orbit taking 3,600 years. If that were true, the Sumerians would have needed a powerful telescope to see it and modern astronomers would be well aware of it. Such prophecies probably say more about human psychology than they do about the future of our planet!

Solar storms Solar activity has an 11-year cycle, so most of us have lived through several cycles unscathed. The present cycle started late and it seems less active than normal. Nonetheless, the Sun does sometimes eject storms of charged particles towards the Earth. These can knock out satellites and cause power surges in electricity lines – but they are not world-ending.

1,050 Ma	3,500 Ma	5,000 Ma
Oceans boiled dry. Water vapour creates Venus-like greenhouse effect	Milky Way starts to collide with Andromeda galaxy, increasing the risks of impacts	Sun swells into a red giant. Earth becomes a lifeless cinder

Terraforming

It is only a matter of time and money before humans set up a base on the Moon and land on Mars. Eventually, thoughts will turn to colonization. Potentially, Mars could be made more Earth-like. This would involve releasing vast quantities of carbon dioxide, perhaps from Martian polar caps or underground reserves, to create an enhanced greenhouse effect and raise temperatures sufficiently for liquid water to exist on the surface. Then, much as on the early Earth, bacteria could be put to work to start producing an oxygen atmosphere. It might take millions of years, but it would give us a second home.

Bombardment There is a real potential threat from asteroid impacts. But, unlike the dinosaurs, humans have powerful telescopes and a space programme. The Spaceguard project is a collection of several international programmes set up to identify any large object that might come close to the Earth. They have so far catalogued more than 1,000 objects larger than 200 metres (656 ft) in size that might come within a range 20 times the distance of the Moon to Earth. On 8 November, 2011, one of them, a dark object 400 metres (1,312 ft) across called YU55, passed safely, as predicted, 324,900 kilometres (201,884 miles) from Earth. The first believed to pose a threat is a 1-kilometre (0.62 mile) object heading for Earth in 2880, by which time there should be a good strategy for deflecting it (see box: Deflecting Armageddon).

Cosmic threats Threats from beyond our solar system are harder to predict. It takes the Sun about 240 million years to orbit the galaxy, during which time it passes through the galaxy's spiral arms. That might stir up long-period comets and increase the risk of bombardment, but it does not seem to correlate with mass extinctions of the past.

Another risk could come from a nearby exploding star. When stars much more massive than our own Sun run out of nuclear fuel they collapse, triggering a supernova explosion. The burst of radiation from one nearby might damage the Earth's ozone layer, but there would be little other damage. More serious might be a nearby hypernova – a stellar explosion so big that it creates a black hole where the star once was and produces an intense jet of gamma rays. Were such a jet directed towards the Earth, the initial blast might cause severe radiation damage across the hemisphere of our planet exposed to it. As the Earth continued to rotate, it might be roasted in radiation like a chicken on a spit. But such a hypernova is extremely rare in an evolved galaxy such as our own.

The expanding Sun The most real and inevitable threat to life on Earth comes from our own Sun. Eventually, the Sun will run out of nuclear fuel at its core. It is too small to explode as a supernova, but it will start to expand and form a red giant star. The bloated mass of incandescent gas will expand to engulf Mercury and Venus. It probably won't reach the Earth, but its heat and a gale of charged particles streaming from it will strip away our atmosphere and boil the oceans dry. The home planet will be left as a burnt-out cinder. The good news is that this is not likely to happen for another four to five billion years.

In search of other worlds In the last decade or so, astronomers have begun to detect the presence of other solar systems. By the end of 2011 nearly 2,000 had been catalogued. Most are revealed by giant planets (like Jupiter or bigger) tugging on their central star, but there is evidence for a handful of Earth-like planets at distances that would make the presence of liquid water possible. Some may already support life, be it bacteria or civilization. Some may be available for colonization.

Such is the vastness of space that, unless some faster-than-light warp drive is invented, it may take thousands of years to reach a new home. Perhaps the colonists will be transported in hibernation. Perhaps generations will be born in transit. Perhaps our descendants will have achieved some sort of immortality in body or machine. But once established, intelligence is not going to give up its hold on the galaxy easily.

> **Civilization exists by geological consent, subject to change without notice.**
>
> **Will Durant, 1935**

the condensed idea
Reach for the stars!

Glossary

Accretion The process by which small particles merge together into larger particles, eventually building up into planets.

Asthenosphere This is the softest layer of the mantle, just beneath the lithosphere. Though mostly solid, it moves with mantle convection, carrying the tectonic plates with it. Because it is hot and soft, seismic waves travel through it at low velocities.

Basalt Fine-grained, dark-coloured volcanic lava created by partial melting in the upper mantle. It is the most common volcanic rock and makes up most of the ocean crust and the outpourings of shield volcanoes.

Continent One of the seven large land masses on the Earth's surface, made up of rocks that are thicker and less dense than those of the ocean crust.

Crust The thin skin of rock that covers the Earth's surface. The ocean crust is on average 7 kilometres (4½ miles) thick, but continental crust can be between 20 and 60 kilometres (12½ and 37¼ miles) thick.

Earthquake The sometimes violent shaking of the ground due to movement along a fault. Earthquakes are most frequent close to the boundaries of tectonic plates.

Epicentre The point on the Earth's surface directly above an earthquake's hypocentre or focus, where the ground fractures.

Erosion The process by which rocks wear away either physically, often due to water, wind or ice, or chemically, often due to water acidified by dissolved carbon dioxide.

Fault A crack in the Earth's crust across which there is relative motion of the rocks, usually in an earthquake. The motion can be horizontal (a strike-slip fault) or largely vertical (a dip-slip fault). The fault plane can be angled to the vertical such that, under extension, it is the overlying block that moves downwards (a normal fault). Or, under compression, the upper block can move upwards (a reverse fault). A reverse fault at an angle of less than 45 degrees is known as a thrust fault.

Fold An undulating deformation in layered rock due to movements in the Earth's crust. A fold that is arched upwards is known as an anticline; a basin-shaped fold is a syncline. In regions of high deformation such as the Alps, over-folds or nappes can develop.

Fossil The traces of prehistoric plants or animals preserved within rocks. Fossils can contain material from the original organism, or material can be replaced by minerals. Trace fossils include marks such as burrows and footprints left by the organism during life.

Fossil fuel A carbon-rich fuel produced by the decay, burial and fossilization of organic remains. Fossil fuels include coal, oil and natural gas. They take millions of years to form but are now being used up within decades.

Ga Billions of years.

Gondwana (formerly Gondwanaland) The great southern continent that resulted from the break-up of a supercontinent about 540 million years ago. It included present-day Antarctica, Australia, India, South America and Africa. Two hundred million years later, it merged with northern land masses to form the supercontinent of Pangaea.

Granite An abundant igneous rock formed by melting deep within a continent. Granite can rise through the crust to form large, domed structures called batholiths. It cools slowly and as a result contains large crystals of feldspar, quartz and mica.

Greenhouse effect The warming of a planetary surface due to gases such as water vapour, carbon dioxide and methane, which allow sunlight in but prevent heat from escaping. Without the greenhouse effect the Earth would be frozen over, but increasing levels of greenhouse gases are now causing excess warming.

Igneous rocks Rocks formed from molten magma. There are two main types: extrusive igneous rocks, which spew out onto the surface of the Earth through fissures and volcanoes; and intrusive igneous rocks such as granite, which push up beneath other layers.

Isotope Atoms of the same element but with different atomic weights due to having different numbers of neutrons in their nuclei. Some isotopes are radioactive and decay with well-known lifetimes. Modern mass spectrometers can measure the ratios of different isotopes with amazing accuracy, using them to date rocks or reveal the processes involved in their formation.

ka Thousands of years.

Laurasia The northern component of the supercontinent of Pangaea. It separated from the southern continent of Gondwana about 200 million years ago. It included Europe, North America and most of Asia.

Lava Magma that has erupted on the surface of the Earth.

Lithosphere The hard, brittle region that includes the Earth's crust and the top of the mantle. Together, these form lithospheric slabs or plates, the principal players in continental drift. The lithosphere is no thicker than the crust along mid-ocean ridges, but older ocean lithosphere can be 100 kilometres (62 miles) thick. Beneath continents, the lithosphere can reach thicknesses of more than 200 kilometres (124 miles).

Ma Millions of years.

Magma Molten rock derived, normally, from partial melting in the upper mantle. Sometimes it remains within magma chambers underground, but it can also erupt on the surface, in which case it is called lava.

Magnetosphere The magnetic envelope produced by the Earth's magnetic field and reaching far into space. It traps the Van Allen radiation belts above the Earth, but shields the planet from the stream of charged particles blowing out from the Sun.

Mantle The region of the Earth from the base of the crust to the top of the core, 2,900 kilometres (1,800 miles) down. It is composed of high-density forms of silicate rock. There is a clear boundary about 670 kilometres (416 miles) down between the upper and lower mantle. This may represent a difference in composition or just in density. Convection in the mantle drives plate tectonics.

Mantle plume A column of low-density hot rock slowly rising through the Earth's mantle. A mantle plume may reach all the way down to the core–mantle boundary and represents the principal route for heat to convect from the Earth's interior. The top of a mantle plume is often marked by volcanic activity.

Metamorphic rocks Rocks that have been transformed by heat or pressure. They can form from both sedimentary and igneous rocks and vary from lightly metamorphosed pilites to extremely altered gneisses.

Mid-ocean ridge A system of ridges running down the centres of oceans, along which new ocean crust is forming. The ridge is normally bisected by a rift within which new crust is created. Sometimes the ridge is offset by transform faults perpendicular to it.

Moho or Mohorovičić discontinuity A distinct layer that marks the base of the Earth's crust. Although the mantle lithosphere beneath is all part of the same tectonic plate, compositional differences mean that the Moho reflects seismic waves.

El Niño, La Niña El Niño is the name given to the warm ocean current in the Pacific that sometimes flows east towards the coast of South America around Christmas, disrupting fisheries and bringing storms and floods to the Americas but drought to the Western Pacific. It is sometimes followed a year later by La Nina, a cold current with opposite effects.

Pangaea The most recent super-continent completed its formation about 320 million years ago in the Carboniferous period. It split apart again about 200 million years ago. The continents may come together once again to form Pangaea Ultima in another 200 million years or so.

Plate tectonics The mechanism by which slabs or plates of the Earth's lithosphere can move relative to one another in a process that combines seafloor spreading and continental drift.

Seafloor spreading The process by which new ocean crust is formed and moves outwards at mid-ocean ridges.

Sedimentary rocks Rocks composed of material that has been physically or chemically eroded from other rocks. They can be deposited at land or sea, but most are marine and built up in layers.

Seismic waves Vibrations running through rocks, they can take the form of shear waves or pressure waves and are reflected or refracted by different layers within the Earth much as light is by a lens. They are produced by earthquakes or by artificial explosions and are used by geologists to prospect for oil or probe the deep structure of the Earth.

Subduction The process by which old cold ocean lithosphere sinks back down into the Earth's mantle in an ocean trench or beneath a continental margin. The process is often accompanied by earthquakes and volcanic activity.

Tsunami An ocean wave created by a submarine earthquake or landslide. Tsunamis can cross an entire ocean, building up into a destructive wall of water as they approach a coastline and shallow water.

Unconformity A boundary between layers, usually of sedimentary rock, that marks a break in deposition. Sometimes, the older layers beneath will have been tilted or folded and eroded before fresh layers form above the unconformity. Unconformities helped James Hutton to understand the deep time of geology.

Volcano An eruption of extrusive igneous rocks onto the Earth's surface. It might be no more than a fissure, or it could build into a high mountain topped by an active crater. Sometimes after an eruption a volcano will subside back to form a circular caldera.

Index

Quercus Editions Ltd
55 Baker Street
7th floor, south block
London
W1U 8EW

First published in 2012

Copyright © 2012 Martin Redfern

The moral right of Martin Redfern to be identified as the author of this work has been asserted in accordance with the Copyright, Design and Patents Act, 1988.

A catalogue record of this book is available from the British Library

UK and associated territories: ISBN 978 1 78087 161 5
Canada: ISBN 978 1 84866 196 7

Printed and bound in China

10 9 8 7 6 5 4 3 2 1

Picture credits
All images drawn by Patrick Nugent, except pages 14, 55, 158, 175, 197 by Pikaia Imaging, and page 166 by Jon Hughes and Russell Gooday of Pixel-shack.com.

Acknowledgements

This book would not exist without the help and patience of many people. I would like to thank the many geologists who have given so freely of their time and understanding over the years. I would like to thank Bridget Walton and Kirsten Dwight for understanding what I'm trying to say even when it doesn't turn out like that, and Ted Nield for correcting some of my worst geological mistakes. Finally I would like to thank Slav Todorov and Amy Visram at Quercus for an original format and intelligent editing. Any remaining errors are down to me. But don't worry unduly about the facts and figures. This is a book about ideas; ideas that are bigger than us all.

Martin Redfern, January 2012